Towards Healthy and Sustainable Human Settlement: The Ecological and Cultural Connation of Landsenses

Towards Healthy and Sustainable Human Settlement: The Ecological and Cultural Connation of Landsenses

Guest Editors

Jiang Liu
Yuhan Shao
Xin-Chen Hong

MDPI

Basel • Beijing • Wuhan • Barcelona • Belgrade • Novi Sad • Cluj • Manchester

Guest Editors

Jiang Liu
School of Architecture and
Urban-Rural Planning
Fuzhou University
Fuzhou
China

Yuhan Shao
College of Architecture and
Urban Planning
Tongji University
Shanghai
China

Xin-Chen Hong
School of Architecture and
Urban-Rural Planning
Fuzhou University
Fuzhou
China

Editorial Office
MDPI AG
Grosspeteranlage 5
4052 Basel, Switzerland

This is a reprint of the Special Issue, published open access by the journal *Land* (ISSN 2073-445X), freely accessible at: https://www.mdpi.com/journal/land/special_issues/50SY660Q96.

For citation purposes, cite each article independently as indicated on the article page online and as indicated below:

Lastname, A.A.; Lastname, B.B. Article Title. *Journal Name* **Year**, *Volume Number*, Page Range.

ISBN 978-3-7258-6820-9 (Hbk)
ISBN 978-3-7258-6821-6 (PDF)
https://doi.org/10.3390/books978-3-7258-6821-6

Contents

land

MDPI

Editorial

Towards Healthy and Sustainable Human Settlement: The Ecological and Cultural Connation of Landsenses

Jiang Liu [1,*], Yuhan Shao [2] and Xinchen Hong [1]

1 School of Architecture and Urban-Rural Planning, Fuzhou University, Fuzhou 350108, China; xch.hung@outlook.com
2 College of Architecture and Urban Planning, Tongji University, Shanghai 200438, China; shaoyuhan@tongji.edu.cn
* Correspondence: jiang.liu@fzu.edu.cn

The construction of a healthy and sustainable living environment requires a systematic integration of multidimensional elements such as the environment, society, and economy. Among these, cultural factors profoundly influence people's cognition and value orientation towards landscapes, while ecological factors are directly related to the health status and sustainability of ecosystems themselves. Achieving a deep integration of cultural identity inheritance and the maintenance of ecological functions is not only crucial for preserving the long-term resilience and stability of ecosystems, but it also significantly impacts the sustainability of human well-being [1]. In the process of landscape cognition and utilization, people's behaviors play a significant role in the intrinsic ecological relationships and cultural connections of the landscape. Landscape perception, which includes sensory experiences generated through human–environment interactions, encompasses various dimensions such as vision, hearing, smell, taste, and touch. This provides an important perspective for understanding the underlying mechanisms of human–environment relationships. A systematic understanding and respect for landscape perception is an effective way to optimize land resource protection and utilization strategies, as well as to promote harmonious coexistence between humans and nature. Therefore, incorporating landscape perception into the design and planning framework of human settlements holds significant theoretical and practical importance.

Current research frontiers have profoundly revealed a paradigm shift in landscape perception. Environmental neuroscience studies indicate that landscape perception is not merely a simple aggregation of single sensory information, but rather a multisensory interactive process that relies on the neural cognitive system to filter, integrate, and enhance information [2]. This has propelled landscape perception from a singular aesthetic evaluation towards a systematic analysis that integrates dimensions such as ecology, culture, and health [3]. This integrative perspective demonstrates significant practical implications in key areas. For instance, in terms of ecological resilience, the "four-dimensional capabilities" framework of hydrological ecological infrastructure (regulation, buffering, carrying, and symbiosis) provides important theoretical support and implementation pathways for systematic sustainable development [4]. In terms of cultural perception, the value recognition of heritage landscape conservation depends on the perception of "historical authenticity and the spirit of place" [5], while quantitative assessments of cultural ecosystem services in urban ecosystems have strongly evidenced the crucial role of cultural value perception in driving conservation behaviors [6]. Regarding health promotion mechanisms, research shows that the objective characteristics of street environments, such as green view ratios, have a dominant influence on running behavior [7]. It is noteworthy that merely increasing the quantity of urban green space does not significantly improve residents' mental health [8]; instead, integrating multisensory experience design can effectively enhance the restorative benefits of environmental perception, thereby improving the well-being satisfaction of vulnerable groups [9]. This underscores the importance of multidimensional perceptual interactions.

The theory of landsense ecology breaks through the traditional research paradigm of ecology, using ecological principles to expand multidimensional perceptual interactions into a systematic integration of natural elements, physical senses, psychological perceptions, and socio-economic factors [10]. Empirical research indicates that in typical "landsense carriers" such as urban parks, visual and tactile perceptions significantly influence visitor satisfaction [11], while auditory and olfactory perceptions are also indispensable factors [12]. This further emphasizes that the design of "landsense carriers" effectively promotes public resonance with sustainable visions through the collaborative action of multiple senses [11]. This deep integration of multidimensional perception not only enhances individual health and well-being (such as stress relief and social interaction) but also optimizes ecosystem service functions through "landsense creation," achieving a harmonious coexistence between humans and the natural living community.

The aforementioned advancements point to a core consensus: constructing a healthy and sustainable living environment requires the integration of cross-scale and multi-entity landscape perception, supported by a theoretical framework built from multidisciplinary perspectives, to promote the deep coordination of ecological processes and cultural practices. Against this backdrop, to systematically address this complex challenge that spans multiple scales and dimensions, this Special Issue focuses on the theme "Towards a Healthy and Sustainable Living Environment: The Ecological and Cultural Connation of Landsenses." It centers on five major research themes—environmental governance, cultural preservation, health promotion, resilient/restorative landscapes, and community engagement—aiming to provide interdisciplinary theoretical support and practical pathways.

This Special Issue includes a total of eight research papers, primarily exploring the following three core topics:

The Multidimensional Value of Ecological and Cultural Integration

The deep integration of ecology and culture is fundamental to constructing a healthy living environment. Multisensory landscape design, through the interaction of visual, auditory, and olfactory stimuli, can significantly enhance the health benefits of public spaces (List of Contributions 1,2). Research indicates that a sound pressure level of 77 decibels is a critical threshold for emotional perception in urban parks; exceeding this threshold can significantly reduce feelings of pleasure while increasing feelings of calmness (List of Contributions 3). This finding not only reveals the complex impact of the sound environment on human emotional responses but also provides a scientific basis for the design of soundscapes in urban parks. Furthermore, biophilic design models confirm that the cumulative effects of multisensory natural experiences on psychological restoration arise from the synergistic interactions of multiple senses, rather than from a linear accumulation of individual sensory stimuli. Combinations of dual sensory interactions—such as visual–auditory, visual–thermal, visual–olfactory, auditory–thermal, and auditory–olfactory—can significantly enhance the overall perception of a space, alleviating and improving people's emotions (List of Contributions 2).

The ecological wisdom of historical landscapes offers valuable lessons for modern design. The "mountain–water–city" pattern of West Lake, along with its four pillars of historical wisdom—holistic approaches, ecological technology, dynamic management, and landscape aesthetics—has not only optimized ecological functions but also carries profound historical and cultural significance, providing considerable insights for contemporary lake landscape construction (List of Contributions 4). This reflects the dynamic balance between ecological processes and cultural memory, offering theoretical and practical support for building a healthy and sustainable living environment.

The Synergistic Mechanism of Cultural Heritage Preservation and Community Participation

The vibrant transmission of cultural heritage relies on livelihood capital and community involvement. Farmers' awareness of natural and cultural capital is significantly related to their willingness to protect these assets. In rural heritage sites, farmers serve as the direct guardians and transmitters of cultural heritage. Their methods of utilizing natural capital and their understanding of cultural capital directly influence the preservation status of cultural heritage (List of Contributions 5). Community residents are more likely to actively engage in the protection of cultural heritage only when they genuinely recognize its value and derive tangible benefits from it (List of Contributions 6). This underscores the critical role of community participation in sustainable development. Moreover, the protection of cultural heritage must extend beyond the maintenance of physical spaces. It is essential to construct a composite governance system that integrates "ecological, cultural, and economic" dimensions through economic incentives and local identity. This approach fosters a holistic framework for ensuring the sustainable preservation of cultural heritage while enhancing community resilience and engagement.

Technological Innovation and Interdisciplinary Integration

The introduction of multi-source data and quantitative tools offers a new paradigm for research in human living environment construction. The Partial Least Squares Structural Equation Modeling (PLS-SEM) model accurately reveals the mechanisms by which landscape features influence human behavior, as well as the significant mediating role of multisensory perception between street space characteristics and public behavior (List of Contributions 7). Moreover, multi-source big data analysis can integrate data resources from various channels and formats, utilizing techniques such as data mining to explore the catalytic effects of landscapes on the potential benefits for broader regional development (List of Contributions 8). Together, these approaches illuminate the complex value of landscapes from both micro-individual and macro-system perspectives. The intersection of landscape history and environmental psychology provides a deeper understanding of how landscape elements affect human psychological perception and emotional experiences, further contributing to the theoretical foundation of restorative landscape design (List of Contributions 1,4).

To realize the vision of a healthy and sustainable living environment, this Special Issue calls for innovative technological methods that integrate advanced technologies such as Multi-sensory Public Participation GIS (MSPPGIS) and machine learning [13]. These technologies can dynamically respond to the spatiotemporal heterogeneity of cultural ecosystem service (CES) needs among different social groups, enabling more precise quantification of how landscape perception drives ecological maintenance and cultural practice behaviors.

Furthermore, this Special Issue aims to inspire broader and deeper academic discussions and practical explorations. We sincerely welcome more scholars to join this field of research to jointly promote theoretical innovation and practical application, injecting sustained momentum into the construction of healthy and resilient living environments, and advancing global research on human living environments to a higher developmental stage.

Author Contributions: Conceptualization, writing—Original draft preparation, J.L.; writing—Review and editing, Y.S., X.H.; funding acquisition, J.L. All authors have read and agreed to the published version of the manuscript.

Funding: This research was funded by the National Natural Science Foundation of China (52378049).

Conflicts of Interest: The authors declare no conflicts of interest.

List of Contributions:

1. Dong, W.; Dai, D.; Shen, P.; Zhang, R.; Liu, M. How Public Urban Space Enhance Restoration Benefits Through Combined Multisensory Effects: A Systematic Review. *Land* **2024**, *13*, 2018.
2. Yin, J.; Zhu, H.; Yuan, J. Health Impacts of Biophilic Design from a Multisensory Interaction Perspective: Empirical Evidence, Research Designs, and Future Directions. *Land* **2024**, *13*, 1448.
3. Jin, T.; Lu, J.; Shao, Y. Exploring the Impact of Visual and Aural Elements in Urban Parks on Human Behavior and Emotional Responses. *Land* **2024**, *13*, 1468.
4. Jia, L.; Lin, Q.; Wang, X.; et al. Wisdom of Landscape Construction of China's West Lakes in Historical Period and Its Implications. *Land* **2025**, *14*, 18.
5. Li, S.; Cheng, Y.; Cai, J.; Zhang, X. Influence of Livelihood Capitals on Landscape Service Cognition and Behavioral Intentions in Rural Heritage Sites. *Land*. **2024**, *13*, 1770.
6. Ladičorbić, M.M.; Dragin, A.S.; et al. Towards Healthy and Sustainable Human Settlement: Understanding How Local Communities Perceive and Engage with Spa Tourism Development Initiatives in Rural Areas. *Land* **2024**, *13*, 1817.
7. Han, T.; Tang, L.; Liu, J.; Jiang, S.; Yan, J. The Influence of Multi-Sensory Perception on Public Activity in Urban Street Spaces: An Empirical Study Grounded in Landsenses Ecology. *Land* **2025**, *14*, 50.
8. Yin, Y.; Ma, D.; Xu, X. The Landscape Catalytic Effect of Urban Waterfronts—A Case Study of the Huangpu River in Shanghai. *Land* **2025**, *14*, 422.

References

1. Thomas, H.; Elmqvist, T.; Frantzeskaki, N.; McPhearson, T.; Olsson, P.; Gaffney, O.; Takeuchi, K. Sustainability and resilience for transformation in the urban century. *Nat. Sustain.* **2019**, *2*, 267–273. [CrossRef]
2. Chen, Z.; Liu, J.; Lin, G. From Environmental Neuroscience to Multisensory Landscape Perception. *Landsc. Archit. Front.* **2024**, *12*, 4–6.
3. Zheng, T.; Pan, Q.; Zhang, X.; Wang, C.; Yan, Y.; Van De Voorde, T. Research Note: Linking sensory perceptions with landscape elements through a combined approach based on prior knowledge and machine learning. *Landsc. Urban Plan.* **2024**, *242*, 13. [CrossRef]
4. Yu, K. Toward Resilient Hydro-Ecological Infrastructure: A "Four Capacities" Framework for Sustainable Development. *Landsc. Archit. Front.* **2025**, *13*, 5–11.
5. Li, Y.; Cao, L.; Gao, W. Perceptual dimensions of heritage canal landscapes and their impact on conservation value cognition. *Ecol. Indic.* **2025**, *176*, 113626. [CrossRef]
6. Krsnik, G.; Illán-Fernández, E.J. Assessing indicators and preferences of cultural ecosystem services in urban areas: A case study of Murcia, Spain. *Landsc. Ecol.* **2024**, *39*, 190. [CrossRef]
7. Dong, L.; Jiang, H.; Li, W.; Qiu, B.; Wang, H.; Qiu, W. Assessing impacts of objective features and subjective perceptions of street environment on running amount: A case study of boston. *Landsc. Urban Plan.* **2023**, *235*, 104756. [CrossRef]
8. Ha, J.; Kim, H.J.; With, K.A. Urban green space alone is not enough: A landscape analysis linking the spatial distribution of urban green space to mental health in the city of Chicago. *Landsc. Urban Plan.* **2022**, *218*, 104309. [CrossRef]
9. Lin, M.; Lin, X.; Wang, Y. How sensory stimuli and barrier-free environments through restorative environmental perception influence visually impaired Individuals' satisfaction with urban parks. *Landsc. Urban Plan.* **2025**, *256*, 105293. [CrossRef]
10. Zhang, L. Correction: How to measure the impact of landsenses ecology on sustainable development? A review of people-oriented emerging approaches. *Environ. Sci. Eur.* **2024**, *36*, 190. [CrossRef]
11. Zhao, J.; Yan, Y.; Deng, H.; Liu, G.; Dai, L.; Tang, L.; Shi, L.; Shao, G. Remarks about landsenses ecology and ecosystem services. *Int. J. Sustain. Dev. World Ecol.* **2020**, *27*, 196–201. [CrossRef]
12. Zheng, T.; Yan, Y.; Lu, H.; Pan, Q.; Zhu, J.; Wang, C.; Zhang, W.; Rong, Y.; Zhan, Y. Visitors' perception based on five physical senses on ecosystem services of urban parks from the perspective of landsenses ecology. *Int. J. Sustain. Dev. World Ecol.* **2020**, *27*, 214–223. [CrossRef]

13. Korpilo, S.; Nyberg, E.; Vierikko, K.; Nieminen, H.; Arciniegas, G.; Raymond, C.M. Developing a multi-sensory public participation GIS (MSPPGIS) method for integrating landscape values and soundscapes of urban green infrastructure. *Landsc. Urban Plan.* **2023**, *230*, 104617. [CrossRef]

land

MDPI

Review

Health Impacts of Biophilic Design from a Multisensory Interaction Perspective: Empirical Evidence, Research Designs, and Future Directions

Jie Yin [1,2,*], Haoyue Zhu [1] and Jing Yuan [3,*]

1 College of Architecture and Urban Planning, Tongji University, Shanghai 200092, China; zhuhaoyue0418@tongji.edu.cn
2 Key Laboratory of Ecology and Energy-Saving Study of Dense Habitat (Ministry of Education), Tongji University, Shanghai 200092, China
3 Center for Built Environment, University of California, Berkeley, CA 94607, USA
* Correspondence: jieyin@tongji.edu.cn (J.Y.); jingyuan10@hotmail.com (J.Y.)

Abstract: Biophilic design introduces a variety of sensory elements into the built environment, incorporating a natural experience into daily life. From the environmental psychology aspect, the effects of multisensory stimulus are not a result of simply adding the effects of uni-sensory stimuli, and the interactive effects among sensory approaches need to be considered. With the demand for promoting health and well-being, increasing numbers of studies began to investigate sensory systems other than just the visual cue of the biophilic design. This review focuses on the multisensory effect of biophilic design on human psychological and physiological responses as well as cognitive function. We summarized empirical evidence of the interactions between two and three sensory modalities from existing research. In addition, we systematically summarized the key methods and technologies used in experimental studies that explore the health benefits of biophilic design in terms of study types and population, environmental exposure simulation, health benefit measurement, and experimental process design. We finally identified some knowledge gaps in and future directions for biophilic design studies from the multisensory interaction perspective. The experimental design of quantifying the effects of multisensory interaction and its lasting effects are the focus of future research.

Keywords: multisensory interaction; biophilic design; experimental research; psych-physiological responses; cognitive function

1. Introduction

Biophilia is the term popularized by Harvard biologist Edward O. Wilson to describe human's innate connections with nature [1]. Therefore, biophilic design particularly emphasizes the connection between humans and nature and enhances the comfort and well-being of the built environment by incorporating natural elements and processes. For example, the visual connection with nature could lower blood pressure and heart rate [2] and positively impact attitude and overall happiness [3]. In addition, the non-visual connection with nature could reduce systolic blood pressure and stress hormones [4,5] and result in perceived improvements in mental health and tranquility [6,7]. Yale social ecologist Stephen Kellert proposed the three-dimensional framework for biophilic design, suggesting that "human satisfaction and wellbeing continue to be reliant on perceiving and responding to sensory variability, especially when this occurs in structured and organized ways within the built environment" [8]. Biophilic design becomes a reasonable approach to encourage the multi-sensorial experience of contact with nature in the built environment, which could provide positive effects on human health and well-being.

The perception of the built environment by individuals is inherently multisensory, with the five common senses of the human body being auditory, olfactory, taste, visual, and haptic. However, in studies related to the current built environment, taste is an abstract aspect that is not easily integrated into the environment. Meanwhile, thermal sensation as a passive tactile contact is a key element of haptics, which could seriously affect people's comfort and perception levels in the built environment, therefore many experiments have used thermal sensation as an important indicator. Our review focuses particularly on the four visual, auditory, thermal, and olfactory sensory factors, which are the most influential in built environments[1].

Previous studies have thoroughly examined the independent effects of single-sensory stimulation on human comfort, physiological and psychological responses, and cognitive function. For example, the visual connection with nature could help with mental stress recovery [9] and improve cognitive performance [10]. A view of greenery from windows has been shown in several studies to have restorative effects on people's well-being. Specifically, a landmark study from Ulrich found that patients who had a green view in their wards recovered faster and required less pain medication than those who had a brick wall view [11]. Li and Sullivan also proved that a classroom view of green landscapes leads to better performance in attention and increased recovery [12]. In addition, types of view are crucial for restoration. Felsten pointed out that a view containing water and a more dramatic nature scene is more restorative [13]. Besides the view, daylight is another visual clue of nature that has multiple health advantages [14,15]. For instance, Kullar and Lindsten found that working in a classroom without daylight may disturb the hormone pattern and negatively influence students' academic performance [16]. Zadeh et al. also found that daylight can positively affect circadian rhythms and reduce morning sleepiness [17]. Research on the soundscape in the built environment is always related to the negative effects of noise. For example, studies have shown that the presence of urban man-made noise can affect human memory [18], attenuate attention [19], and increase depressive mood [20]. Meanwhile, natural sound as a positive-effect soundscape can be beneficial to human stress recovery and memory performance. Benfield et al. found that listening to pure natural sound can improve people's memory ability and a 3 min exposure to natural sound following an unsettling video can lead to greater mood recovery [21]. As a noise-masking tool, natural sound can improve speech privacy while increasing cognitive performance, concentration, and worker satisfaction [22]. Thermal comfort in different climate zones has been well studied (Thermal Comfort Database II) [23], including different thermal sensations on different body parts [24–26], physiological models for describing thermal comfort level [27–29], and personal comfort devices for satisfying personal thermal needs [30]. A noticeable change in research topics is that studying the undesirable thermal environment has shifted towards studying the desirable quality of air movement and personal comfort control systems [30,31]. Researchers proposed that comfort standards should rely on dynamic and spatially variable indoor climates for comfort evaluation. The concept of alliesthesia—the physiological basis for thermal delight proposed by De Dear—supports the transformation from neutrality to delight [32], which also has analogies to other senses as well. Olfactory atmospheres in the built environment were shown to have a significant influence on our health and well-being [33]. Sick-building syndrome (SBS) is an office condition defined by the World Health Organization in 1979 and describes the illness caused by air pollution in the built environment. The symptoms include headache, dizziness, nausea, eye, nose/throat irritation, dry cough, and even personality changes. Many of the outbreaks of SBS that have been reported over the last half-century were linked to an unfamiliar and unpleasant smell in the poorly ventilated environment [34,35]. To alleviate SBS symptoms, researchers have demonstrated that both increased ventilation rates and reduced air pollution can exert a significant influence on subjective health and well-being [36]. Meanwhile, undesirable and negatively valenced malodors tend to be associated with negative outcomes on the health, well-being, and mood of those who are exposed to them. Meanwhile, it is important to note that pleasant ambient scents can also be used to help improve our mood and well-being [37]. More recent studies find that short-term exposure to pleasant ambient fragrances, such as grapefruit odor [38], can impact people's attention, anxiety, and mood, with different effect sizes on genders. In terms of the office setting, many researchers have argued that lavender can sometimes be used to enhance productivity by reducing stress [39,40]. These findings indicated that we need to reduce air pollution and increase positive scent stimulus at the same time to promote health and well-being.

Despite this extensive body of work, much of the research remains siloed, with studies often focusing on single-sensory approaches. At the same time, since vision plays a dominant role in our perception [41], other sensory modalities, such as acoustic, olfactory, and thermal senses, are mostly neglected in the design process. For example, when we introduce water features into the indoor environment, not only does the floating water make people relax visually and cool thermally, but also the water sound can mask high-frequency sound and create a relatively quiet space [8]. Therefore, integrating multisensory design can promote the occupants' well-being. For instance, a recent study conducted in a living lab found that biophilic interventions have the most potential to positively mitigate the experience of stressors. Participants felt less stressed in each of the biophilic conditions, especially in the multisensory intervention [42]. However, the effects of multisensory stimuli are not a result of simply adding the effects of uni-sensory stimuli together. Research on multisensory interactions has increased over the last few decades in the field of cognitive neuroscience and neurophysiology. However, the effect of the built environment on human health and comfort through a multisensory approach has not been fully investigated. Only a few studies analyzed different combinations of environmental senses, with less than 9% percent of studies considering visual, acoustical, and thermal senses together [43]. Among existing multisensory studies, most have focused on reducing the negative impacts of different sensory stimuli [44]. However, a higher level of experiential delight is needed to achieve the goal of well-being, we should get beyond mere 'comfort' and design for these higher-level experiences [45]. Specifically, sound privacy, thermal discomfort, low lighting level, and poor indoor air quality (IAQ) have been identified as important factors contributing to an unhealthy environment. Research on beneficial environmental parameters, including natural view, natural sound, scent, aesthetics, security, and ergonomic design factors is still limited.

In summary, research on single senses is now relatively complete, but research on more than two sensory approaches is still limited, this review aims to address this gap by summarizing existing scientific evidence on the perception and response to biophilic environments through various sensory approaches. Additionally, it seeks to identify gaps in the current research and propose directions for future studies, ultimately advocating for a more integrated multisensory approach to design that prioritizes human health and well-being.

2. Methods

We conducted a scoping review following Arksey and O'Malley's framework [46], which is a particularly useful approach when the information on a topic has not been comprehensively reviewed or is complex and diverse [47]. We aim to summarize evidence from existing biophilic design literature, rapidly map the key concepts of biophilia underpinning the multisensory design, and identify gaps in the sensorial evidence. Since this topic involves multiple environmental factors together which have not been reviewed comprehensively before, a scoping review will help us assemble multisensory studies and propose a potential research direction in the biophilic design field (Figure 1).

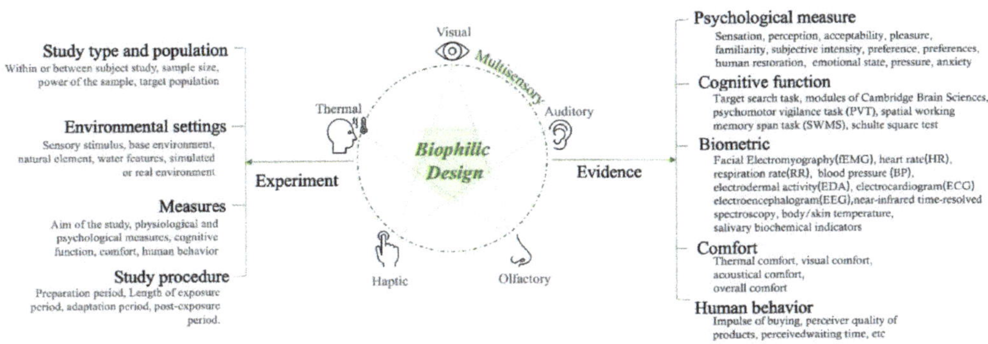

Figure 1. Knowledge framework for multisensory implementation of biophilic design.

First of all, we frame our research question as "What is the current evidence on people's physiological, psychological, and cognitive responses to biophilic design through multisensory approaches?" Specifically, we focused on four well-studied senses in environmental perception studies: visual, acoustical, olfactory, and thermal. Then, we used Google Scholar to search for peer-reviewed papers and book chapters in the fields of building science, public health, and environmental psychology. In terms of the searching process, we started with reviews on single-sensory studies and then explored the experimental studies investigating more than one sensory approach to environmental settings with biophilic features. Initially, a preliminary selection was conducted based on the titles and abstracts of the relevant literature, and 316 research papers were selected. We also use the snowball method to search through the reference list of multisensory studies to increase our database. Since we mainly focused on the positive effects of exposure to biophilic features, we excluded articles evaluating interaction effects caused by negative indoor air quality (IAQ), noise, glare, and dim light in combination with other environmental senses, as well as non-neurophysiological and non-urban scale studies. The detailed inclusion and exclusion criteria are shown in Figure 2. As a result, a total of 26 research papers were identified and included. The details of each paper can be found in the table. We charted the selected studies by key findings, experimental design and procedure, measures, and statistical method. Based on this, we summarized results from different sensory studies and provided a study design methodology for future work.

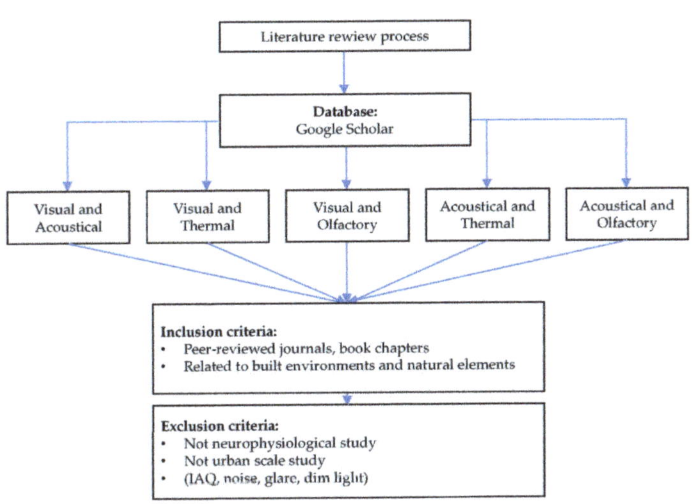

Figure 2. Literature searching criteria.

3. Research Findings

3.1. Bi-Sensory

3.1.1. Visual–Acoustical

Visual–audio interactions can significantly influence the overall perception of a space. This is because when visual and acoustical feelings are coupled, attention paid to the visual cues would change the conscious perception of sound and vice versa. Under the umbrella of biophilic design, the goal of studies involving visual–auditory interactions was to focus on introducing natural scenery and sound into the built environment. Specifically, this topic has been widely explored on the urban scale, where studies found that the pleasantness of the visual environment was related to the judgments of the soundscape [48,49]. Studies also demonstrated that a combined visual–auditory natural environment led to a greater recovery in stress levels than the view-only scenario [50]. In addition, the contribution of natural sound to the overall impression of the urban environment was more significant than natural visual factors when there is a high level of road traffic noise [51].

For the indoor environment, Jahncke et al. found that people prefer offices with natural elements, whereas natural sound was associated with restorative qualities [52]. In addition, a study focusing on open-space offices found that natural sound always had greater effects on people's recovery from fatigue and annoyance compared to environmental noise no matter whether the visual scene was natural or not, and the combination of water and natural environment was the best restorative condition [53]. Based on the visual–auditory interaction, Sun et al. also proposed a personal factor to modify the impact of natural window views on sound perception in a living room, called "audiovisual aptitude" [54]. A recent study conducted in a living lab found that an immersive biophilic indoor environment with visual and auditory biophilic features could improve occupant satisfaction and cognitive performance while reducing stress [42]. Another study found that the visual–auditory environment produced a better acute recovery effect. While with a longer recovery time, the auditory restorative environment might produce the most pronounced stress-recovery effects, followed by the visual restorative environment [55].

Water as a key feature in the biophilic design framework could offer a multisensory experience by providing visual, acoustical, thermal, and tactile delight [8]. Specifically, water sound is a natural and cost-effective tool to mask noise [56–59], reduce physiological stress levels [60], boost cognitive function, and improve mood [3]. However, using water features indoors is rare. Abdalrahman and Galbrun found that the introduction of the water feature which generates a water sound and produces visual stimuli improved perception dramatically [61,62]. They also suggested that water features need to be incorporated into the open-space office with plenty of workstations. More recently, Hasegawa and Lau summarized 27 visual–auditory review studies and highlighted the benefits of using water in an indoor environment, including that water features are influential visual factors for improving psychological responses and audio features have a similar potential [63].

3.1.2. Visual–Thermal

Vision may have a relevant influence on how we perceive other senses. For example, studies found that daylighting may have psychological and physiological interactions with thermal sensation. Kulve et al. found that the changes in visual and thermal comfort were positively related [64]. Specifically, thermal discomfort can be compensated for by lighting color and intensity, resulting in a higher perceived visual quality [64]. Through a controlled experimental investigation by using natural daylighting, Chinazzo et al. summarized that the quantity of daylight could influence thermal perception, resulting in a cross-modal effect, with a low daylight illuminance leading to a less comfortable and less acceptable thermal environment in cold conditions and a more comfortable one in warm conditions [65]. By giving participants more control over daylighting and electrical lighting, Haldi and Robinson concluded that there is no significant interaction between thermal and visual sensation, but people tended to be more thermally and visually comfortable when both conditions met their satisfaction [66]. Whether the relationship is strong or not, we could confidently say that from the psychological aspect, the interaction between daylighting and thermal sensation cannot be ignored. However, previous studies did not find significant interaction effects on physiological parameters, such as skin temperature, cognitive behavior therapy (CBT), and energy expenditure. Although the interaction between daylighting and thermal sensation has been studied for years, another vision-based biophilic component, a window view, was lacking enough literature support. Only one study explored the relationship between window view and thermal comfort and found that people felt significantly cooler when there was an accessible window view [67].

3.1.3. Visual–Olfactory

The combined visual and olfactory stimulation of the environment is relaxing. Song et al. showed that the combination of visual and olfactory stimulation increased sensations associated with "natural" and "realistic" [68]. Li et al. demonstrated that visual–olfactory co-stimulation could reduce fatigue, enhance self-esteem, relieve stress, improve spatial working memory, and reduce psychomotor vigilance errors [69].

3.1.4. Acoustical–Thermal

Previous studies investigated the interaction between acoustical and thermal senses in the built environment, but the results were controversial. Tiller et al. found that acoustic conditions can affect people's thermal comfort with no reverse effect [70]. Pellerin et al. reported that sound level may alter thermal pleasantness only in warm conditions [71]. However, a study in Japan suggested that the thermal environment must be taken into consideration in acoustical studies by claiming that the operating temperature can affect people's acoustical sensations [72]. A more recent study found that acoustic comfort was found to be greater at 25 °C compared to those at 20 °C or 30 °C [73].

As a representative of natural sound, water sound affected not only acoustic perception but also thermal and overall indoor environmental comfort. Yang and Moon [74] used recorded water sound from indoor fountains to study the interaction between sound and thermal comfort and its effect on acoustical perception. They found that water sounds could enhance pleasantness, calmness, and naturalness without increment of noise, loudness, or annoyance in a hot thermal environment.

3.1.5. Acoustical–Olfactory

Olfactory and acoustical elements individually or combined have received the most research attention in the retail context. A study found that when plants' scents and pleasant sounds are combined, consumers rate the environment positively and experience enhanced satisfaction [75,76]. The multisensory effect between scent and sound was also investigated in healthcare research. With cautious usage, ambient scent and music can reduce patients' anxiety in the waiting room of a plastic surgeon [77]. Recently, a study examined the nature of any cross-modal interactions between ambient sound and smell in a laboratory setting designed to capture the sensory cues that might be encountered in a typical urban environment [78]. The influence of odor on sound evaluation showed that the presence of odor had little effect on the evaluations of birdsong and low-volume sound. However, the concentration of odor was associated with a positive evaluation of noise. For example, a study found that the aroma of lilac can reduce the annoyance caused by traffic noise in urban streets [79]. Therefore, a positive sensory stimulus might improve the evaluation of perception through other senses, while a negative sensory stimulus has the opposite effect. In the built environment, acoustical and olfactory elements are highly correlated with each other. For overall comfort, the effect of sound was stronger than odor; while for congruency, the effects of sound and odor were almost equal.

3.1.6. Other Bi-Sensory Elements

Studies on thermal–olfactory and olfactory–touch interactions have been conducted, albeit in a limited number. One study found that thermal discomfort may produce a "revenge effect" for fragrance comfort, and fragrant stimuli resulted in an increase in the beta-band in "slightly warm" environments [80]. Research on the design of medical facilities suggests that when it is challenging to establish direct connections with the natural environment, an indirect natural experience model could be utilized. This may involve incorporating natural materials such as wood to stimulate the sense of touch, evoking olfactory memories with natural scents, and creating a "realistic" natural experience through various means.

3.2. Multisensory Experience

Most of the studies with the multisensory concept only focused on two senses [44]. It may be due to two main reasons: (1) it is hard to control more than two sensory variables in the experimental design; (2) the interactive effects among different sensory experiences might be too small to detect. However, some studies used multiple environmental factors that represent different senses to create a prediction model for indoor environmental comfort. For example, Krüger and Zannin discussed thermal, luminous, and acoustic comfort in the same study and proposed that designers should systematically consider all senses indoors to promote a healthy environment [81]. Huang et al. found that both thermal and acoustical senses dominate the satisfaction level of the indoor environment [82].

There are even fewer publications when we constrain all the existing studies to the realm of biophilic design. Yang and Moon incorporated three different room temperatures and three illuminance levels along with different types and levels of sound (including water sound) [73] and found that the effect of acoustic factors was the greatest on indoor environmental comfort, followed by room temperature and illuminance. Thermal perceptual comfort is susceptible to auditory and visual environments; however, the impact of the visual environment and thermal environment on auditory comfort is not significant [83]. A study investigated the recovery in a sensory-enriched break environment by integrating three sensory stimuli, including natural outdoor view and indoor greenery, bird sound, and herb scent [84]. They concluded that sensory-enriched environments were perceived as more pleasant and restorative compared to the standard break room. High levels of environmental pleasure are associated with low physiological stress responses in the sense of smell, to some extent with auditory stimuli, but not with visual stimuli [85]. Birdsong combined with visual–olfactory stimulation can increase physiological recovery and overall perception of quality assessment but has no significant impact on mental well-being [86]. Zhong et al. demonstrated in a sense walking experiment that high visual and smell environment quality can enhance soundscape evaluations, although the smell environment had a greater impact on the soundscape comfort degree (SCD) than the visual environment in waterfront space in mountainous cities (WSMCs) [87].

4. Study Design Implication

We summarized the research method used by experimental studies focusing on using biophilic elements to generate sensory stimuli for human subjects. Table 1 summarizes the study design of selected studies in detail. They could be divided into four subsections: study type and population, environmental settings, measures, and study procedure. We summarized each subsection with thoughts on the current state of the art, knowledge gaps, and future study design approaches.

Table 1. Summary table of all the experimental studies.

No.	Reference	Sensory				Subject	Tool	Experimental Design	Environmental Procedure	Measures	Main Conclusions
		Visual	Acoustical	Thermal	Smell						
1	Park et al., 2020 [50]	✓[2]	✓			32 ordinary people (16 males, 16 females; age range: 20–39; half of them were in their 20 s and the other half were in their 30 s). Recruited from online study advertisement; hearing Screening; and psychometric Screening; UK	Biometric measures Questionnaire Visual: VR screen (360-degree video) Acoustical: headphones	Visual–auditory interaction vs. visual only Visual: rural landscape (with or without water features) vs. urban scenes Auditory: sound recorded along with the video	1 min baseline clip 1 min stressor clip 1 min recovery clip (experimental exposure) Responding to questionnaire	Psychological response (verbally respond to the questionnaire): Tranquility, preference, and pleasantness for the recovery clips Perceived restorativeness soundscape scale (PRSS) Physiological response: two facial electromyography data points (fEMG), heart rate (HR), respiration rate (RR), and electrodermal activity (EDA)	The rural settings had a better recovery when they were presented as visual–audio combined. Water features led to a greater recovery.
2	Hong and Jeon, 2013 [51]	✓	✓			20 ordinary people (15 males, 5 females; age range: 23–34; M_{age}: 27.2; standard deviation: 2.24). Hearing screening; consistency test; South Korea	Questionnaire Visual: beam projector Acoustical: headphones	Visual-only, audio-only, and visual–auditory interaction Visual: images of streetscapes with a combination of vegetation and water features (photomontage method) Auditory: 9 acoustic stimuli were constructed using 4 individual sounds	Continuous experimental exposure while responding to the questionnaire	Subjective evaluation: Preference for each stimulus Semantic differential test: 12 pairs of adjective attributes (quiet–noisy, calm–loud, pleasant–unpleasant, comfortable–uncomfortable, open–closed, wide–narrow, stable–unstable, harmonious–disharmonious, ordered–disordered, various–monotonous, distinct–ordinary, and natural–artificial.)	Increases in greenery from trees or bushes can improve streetscapes. Bird sounds were more useful for enhancing soundscape quality compared to water. The contribution of acoustic comfort to the overall impression was more significant than visual factors with a high level of road traffic noise.
3	Jahncke et al., 2015 [52]	✓	✓			40 (49 students[3]; 22 males, 27 females; Mage: 24.1). Recruited from the University of Gävle. Hearing Screening; UK	Questionnaire Visual: screen Acoustical: headphones	Visual: open plan office and urban nature Auditory: natural sound, quiet, broadband noise, office noise	Fatigue scenario Control and background questions 1 min exposure to each setting (8 settings) Statement questions Control and background questions	Perceived restorativeness scale (PRS) Restoration likelihood attitude toward the presented setting	Natural sound positively influenced evaluations of the natural setting compared to the office settings. There are significant interactions between acoustic and visual stimuli were found for all measures.
4	Ma and Shu, 2018 [53]	✓	✓			75 (Study1: 30, Study2: 15, Study3: 30; male/femal = 1:1; Mage: 25). Recruited from Tianjin University; hearing screening; working status and stress level assessment; Tianjin, China	Biometric measures Questionnaire Cognitive test Visual: screen Acoustical: headphones	Auditory-only (types and sequences) vs. visual–auditory interaction Visual: open plan office with and without natural elements Auditory: flowing water sound and urban noise	Within-subject study 5 min introduction of stress and attentional fatigue 2 min measure original status 3 min restoration period 2 min measurements of restorative effects 2 min rest (then next experiment unit)	Physiological responses: blood pressure (BP) and heart rate (HR) Psychological experience: 3 emotional states (tension, fatigue, and annoyance) Cognitive performance: task performance	Soundscape elements had an apparent positive effect on tiredness, restoration, and annoyance reduction. Sound elements had a greater effect on psychological restoration compared with visual scenes.
5	Sun et al., 2018 [54]	✓	✓			68 (40 males, 28 Females; Mage = 27.9, SD = 5.05, range: 20–46; 48 obtained a master's degree or higher). Hearing screening; Belgium	Questionnaire Visual: screen Acoustical: speaker	Auditory only vs. visual–auditory interaction Visual: 4 scenarios, airport car, restaurant, aircraft, and city park Auditory: 6 sound groups, the sound was recorded along with the scenario	Part 1: Audio: 3 sound contents Part 2: Video: 3 sound contents (10 min, experiments were repeated for four days for different scenes)	Preference: which of the 3 items sounds most different from the other two?	Audiovisual aptitude may affect the appraisal of the living environment.

Table 1. *Cont.*

No.	Reference	Sensory				Subject	Tool	Experimental Design	Environmental Procedure	Measures	Main Conclusions
		Visual	Acoustical	Thermal	Smell						
6	Abdalrahman and Galbrun, 2020 [62]	✓	✓			31 (16 males, 15 females; range: 24–60, M ¼ 36.3, SD ¼ 9.3). Participants were postgraduate students and staff members of Heriot-Watt University who worked in open-plan offices; hearing screening; UK	Questionnaire Visual: screen Acoustical: headphones	Audio-only vs. audio–visual interaction (Detailed information about Exp 1 is not discussed here). Visual: still images from the animation of the water feature (6 settings) Auditory: water mask: recording of 6 water features (the 20 s each); speech recording: open-plan office	Part 1 (30–35 min) [15 pairs of comparison]: Audio only/visual audio (7 s underwater sound—a 4-step cascade (CA), 1 s silence, and 7 s another underwater sound—a 37-jet fountain (FTW) Part 2 (5–10 min): [6 settings] Audio only/visual audio (7 s unmasked voice, 1 s gap, and 7 s voice masked by water sound) Rate perception changes	Preference of waterscape Sound perception changes	The introduction of a water feature improved the perception of the sound environment and adding visual stimuli improved perception by up to 2.5 times.
7	Galbrun and Calarco, 2014 [61]	✓	✓			38 (19 males, 19 females; range: 24–47; M_{age}: 30.1; standard deviation: 4.47). Hearing screening; consistency test; cultural groups UK	Questionnaire Visual: screen Acoustical: headphones	An audio-only vs. visual-only vs. visual–auditory interaction Visual: photo montages with different water features and the same natural background Auditory: 10 water sounds and road traffic noise	Pair comparison: Audio-only test: a select sound that is more peaceful and relaxing [20 min] + quality analysis of water sound [20—30 min] Visual-only test: the image that prefers to look at [20 min] + rate water feature display [5–10 min] Audiovisual test: feature they prefer in terms of peacefulness and relaxation [20 min]+ rate water feature display [5–10 min]	Pair comparison Sound qualities: semantic assessment, categorization, and evocation Water features' displays as man-made, natural, or neither	Equal attention should be given to the design of both visual and acoustical stimuli. Natural-looking features tended to increase preference scores compared to audio-only paired comparison scores.
8	Liu et al., 2023 [55]	✓	✓			28 (14 males, 14 females). Recruited from Qingdao University; psychiatric screening; climate adaptation screening; unhealthy behaviors such as alcohol and tobacco addiction screening; BMI screening; Qingdao, China	Biometric measures Visual: slides Acoustical: stereophonic loudspeaker	An audio-only vs. visual-only vs. visual–auditory interaction Visual: natural scenes (green trees and forests) Auditory: the sound of naturally running water, with a frequency of 400–500 Hz and a sound level of 40–50 dB	10 min baseline measuring stage 10 min stress induction stage (continuous high-frequency noise) 20 min stress recovery stage (the auditory and the visual-auditory stimuli, respectively)	Continuous electrocardiogram (ECG) data Heart rate variability analysis: mean heart rate, the root mean square of successive differences (RMSSD) between normal heartbeats, the low-frequency and high-frequency power ratio (LF/HF)	The visual and visual–auditory environment produced a better acute recovery effect. In longer recovery time, the auditory restorative environment might produce the most pronounced stress-recovery effects, followed by the visual restorative environment.
9	Aristizabal et al., 2021 [42]	✓	✓			37 (Cohort 1 (6 females, 7 males, M_{age} = 41.85); Cohort 2 (5 females, 8 males, M_{age} = 33.62); Cohort 3 (8 females, 4 males, M_{age} = 33.73) range(18–60);). Hearing and vision screening; cardiovascular disease, psychiatric, stress, depression, drug, and alcohol dependence screening; health assessment; duration of residence screening[4]; Minnesota, USA	Biometric measures Questionnaire Cognitive test Visual: digital screens Acoustical: speakers	An audio-only vs. visual-only vs. visual–auditory interaction Visual: indoor plants and rotating, digital projections of nature including fractal imagery and canopy-type plants Auditory: reminiscent of the natural, regional environment including blowing wind, trickling water, and sounds produced by regional fauna	Pair comparison: Baseline office environment with no environmental aspects (2 weeks) Introducing only visual biological conditions (8 weeks) Introducing only auditory biological conditions (8 weeks) Introducing visual and auditory biological conditions (8 weeks)	Physiological indicators of stress, including changes in heart rate and electrodermal activity Feelings of stress, environmental satisfaction, perceived productivity, mood, and connectedness to nature Objective indicators of cognitive performance Working memory test, inhibition control, and task-switching	Immersive biophilic environments can improve occupants' satisfaction and cognitive performance while reducing stress. Highlight the need to consider non-visual factors in biophilic design.

Table 1. *Cont.*

No.	Reference	Sensory				Subject	Tool	Experimental Design	Environmental Procedure	Measures	Main Conclusions
		Visual	Acoustical	Thermal	Smell						
10	Kulve et al., 2018 [64]	✓		✓		35 (All females; age range: 18–30; M_{age}: 22.2). Recruited via advertisements at the university and the website digi-prik.nl. BMI screening; inclusion criteria: Caucasian females, generally healthy. Exclusion criteria: color blindness, ocular pathologies, medication use, pregnancy, hypertension, general feeling of illness on the day of the experiment, (history of) cardiovascular diseases, and contraindication of the telemetric pill. The Netherlands	Biometric measures Questionnaire Visual: luminance level and color temperature (electrical lighting) Thermal: room temperature	Visual: Study 1: Dim (5 lux) and Bright (1200 lux) with constant color temperature (4000 K) Study 2: Color temperature (2700 K and 5800 K) with constant luminance level (50 lux) Thermal: baseline temperature 29 °C, low temperature 26 °C, and high temperature 32 °C	For both studies: 30 min baseline measure (29 °C) 15 min break 75 min 1st block (26 or 32 °C) 15 min break 75 min 1st block (29 °C) 15 min break 75 min 1st block (26 or 32 °C)	Questionnaire for thermal and visual perception: Thermal: "thermal comfort", "thermal sensation", "preferred temperature change", "self-assessed shivering", and "self-assessed sweating" Visual: "perceived light intensity", "perceived light color", "visual comfort", "preferred light intensity change", and "preferred light color change" Body temperature, skin temperature: 26 body sites; core temperature Energy expenditure: oxygen consumption and carbon dioxide production	Visual perception and thermal perception affect each other. Higher visual comfort levels were correlated with higher thermal comfort votes. Thermal discomfort can be partly compensated by lighting that results in a higher perceived visual comfort.
11	Chinazzo et al., 2019 [65]	✓		✓		84 (42 males, 42 females age range: 18–30). Recruited from the subject pool of the Universities. Vision screening; inclusion criteria (age 18–30 years old, no abuse of alcohol or use of drugs, full-color vision, generally healthy, French-speaking–mother tongue or C2 level, BMI between 18 and 25 kg/m^2, no visual or motor abnormalities). Lausanne, Switzerland	Biometric measures Questionnaire Visual: luminance level (daylighting) Thermal: room temperature	Visual: daylight illuminance (low ~130 lx, medium ~600 lx, and high ~1400 lx)—change filter Thermal: 3 temperature levels (19, 23, and 27 °C) (each participant experiences one T)	For each room temperature 45 min pre-test phase 10 min break, change filter 30 min daylight exposure 1 10 min break, change filter 30 min daylight exposure 2 10 min break, change filter 30 min daylight exposure 3	Subjective perception ratings: 4 types of thermal perception: thermal state (thermal sensation, comfort, and preference) and thermal ambiance (thermal acceptability). and overall perception Evaluate overall comfort The reason for the discomfort Physiological measurements: skin temperature	Cross-modal effects of daylight on thermal responses occurred, but only at a psychological level rather than at a physiological one. Daylight affected only thermal evaluations, not thermal sensation.
12	Ko et al., 2020 [67]	✓		✓		86 (43 males, 43 females; age range: >18). Recruited from the University of California. Vision screening; sleep disorders screening; California, USA	Biometric measures Questionnaire Cognitive test Visual: view content (natural view) Thermal: room temperature	Visual: with or without a window view Thermal: 28 °C (slightly warm condition)	For each window setting: 15 min setup 5 min survey 10 min creativity tests 5 min break in chamber 25 min cognitive tests 5 min survey 10 min break in the reception area	Thermal perception: thermal sensation, comfort, acceptability, and pleasure Mean skin temperature Emotion: circumplex model Cognitive performance: working memory, concentration, short-term memory, spatial planning, and creativity performance test Eye symptoms and perceived stress level	People close to a window can tolerate small thermal comfort deviations. Window view can enhance positive emotions, reduce negative emotions, and improve workers' productivity.

Table 1. *Cont.*

No.	Reference	Sensory				Subject	Tool	Experimental Design	Environmental Procedure	Measures	Main Conclusions
		Visual	Acoustical	Thermal	Smell						
13	Song et al., 2019 [68]	✓			✓	21 (All females; M_{age}: 21.1 ± 1.0 years). Recruited from a Japanese university. Exclusion criteria: smoking, treatment of diseases, menstrual period; Japan	Biometric measures Questionnaire Visual: screen Scent: essential oil odor bag diffuser	Visual-only vs. olfactory-only vs. visual–olfactory interaction Visual: a photograph view of a forest landscape of Hinoki cypress trees (Chamaecyparis obtusa, a type of conifer) Olfactory: hinoki cypress leaf oil	The participant remained sitting, and her physiological responses were continually measured. 60 s (the rest period) viewed a gray Background 90 s (stimuli) the visual, olfactory, combined visual and olfactory Subjective indices evaluation	Near-infrared time-resolved spectroscopy (oxy-Hb concentration in the participant's left and right prefrontal cortex) Heart rate variability and heart rate The HF component of HRV reflects parasympathetic nervous activity and the ratio of LF to HF reflects sympathetic nervous activity Modified semantic differential method	The forest-related stimuli induced a significant decrease in the oxy-Hb concentration in the prefrontal cortex and a significant decrease in sympathetic nervous activity. Significant increases in subjective feelings related to the terms "comfortable", "relaxed", "natural", and "realistic". The combined visual and olfactory stimuli demonstrated an additive effect.
14	Li et al., 2024 [69]	✓			✓	48 (24 males, 24 females; M_{age}: 22.66 ± 1.82). Recruited from college. Vision and olfaction screening; no prior history of mental, cardiovascular, or allergic diseases; anxiety and depression screening; BMI. No significant differences among the four groups in terms of gender ratio, age, height, weight, or body mass index (BMI); Beijing, China	Biometric measures Questionnaire Cognitive test Visual: plant Scent: plant breathing mask	Visual-only vs. olfactory-only vs. visual–olfactory interaction Visual: 2 (plant present vs. absent)—Coriander Olfactory: 2 (scent present vs. absent)—Coriander scent	All tests were in the same period of time (14:00–16:00) 5 min rest (baseline values): completed saliva collection, self-reported questionnaire, and cognitive tests 30 min (stimulation) 5 min rest: completed saliva collection, and self-reported questionnaire 30 min (stimulation) 5 min rest: completed saliva collection, self-reported questionnaire, and cognitive tests	Psychological indicators: the Profile of Mood States (POMS) questionnaire; Electrophysiological indicators: electrocardiogram (ECG), electrodermal activity (EDA), and electroencephalogram (EEG); Salivary biochemical indicators: salivary stress marker (cortisol), proinflammatory cytokines, and untargeted metabolomics; Cognitive performance: psychomotor vigilance task (PVT) and spatial working memory span task (SWMS).	The various types of sensory stimuli associated with coriander plants exhibited different intervention effects on mood and cognition. The combined stimulus demonstrated better effects compared to the single-sensory stimulus. All three stimuli—visual, olfactory, and combined—induced spontaneous neural oscillations associated with relaxation or cognitive function, and significant changes occurred in metabolic pathways related to antidepressant, anti-inflammatory, or neuroprotective effects. It appeared that visual stimuli elicited a greater response from the nervous system, while olfactory stimuli elicited a greater response from the endocrine and immune systems.
15	Yang and Moon, 2019 [74]		✓	✓		54 (25 males, 29 females; M_{age}: 22 ± 1.9). Recruited from university. Hearing screening; South Korea	Questionnaire Acoustical: water sound (speaker) Thermal: room temperature	Acoustical: 2 types and 4 levels of water sound (45, 50, 55 dBA, and 60 dBA) Thermal: 18 °C (cool), 24 °C (neutral), and 30 °C (warm)	For each room temperature: 30 min thermal adaptation 30 min experimental period [25 s sound stimulus + 15 s response (36 sound stimulus)]	Negative acoustic attributes and positive acoustic attributes Acoustic comfort, thermal comfort, and overall comfort	Room temperature affected both thermal perception and acoustic perception. Water sounds affected not only acoustic perception but also thermal and overall indoor environmental comfort.
16	Mattila and Wirtz, 2001 [75]		✓		✓	247 (Female: 75%, less than 20 years old: 65%). Subjects were anyone entering the store who agreed and completed the questionnaire; USA	Questionnaire Acoustical: sound system Scent: diffuser	Acoustical: 3 categories (no music/low arousal music/high arousal music) Olfactory: 3 categories (no scent/low arousal scent/high arousal scent)—Lavender (low arousal) and Grapefruit (high arousal)	In retail Pretest of scent and sound 15 min pre-scent of the store 3 shifts (10:30 a.m.–12:30 pm, 2:00 p.m.–4.00 p.m., and 5:00 p.m.–7:00 p.m.) Randomly select customers leaving the store	Emotional response: arousal dimension and pleasure dimension Approach–avoidance behavioral responses The extent of impulse buying Environment evaluation	When ambient scent and music are congruent with each other in terms of their arousing qualities, consumers rate the environment significantly more positive.
17	Fenko and Loock, 2014 [77]		✓		✓	117 (28 males, 89 females; M_{age}: 47.92; ranged: 14-88). Recruited from the patients of plastic surgeon Dr. Abdul Yousef at the Elizabeth Hospital in Recklinghausen (Germany). Germany	Questionnaire Acoustical: CD player Scent: diffuser	Acoustical: 2 (music present vs. absent)—instrumental music with nature sounds Olfactory: 2 (scent present vs. absent)—Lavender scent	In the waiting room of a German plastic surgeon: Pretest of scent and sound Before the appointment: the demographic questions, evaluation of anxiety, and waiting environment After the appointment: objective and perceived waiting time and manipulation check questions about perceived scent and music	The level of anxiety (Clinical Anxiety Scale and STAI) Evaluation of the waiting environment (Physical Environment Quality Scale) Perceived waiting time duration Objective waiting time	When used separately, each of the environmental factors, music and scent, significantly reduced the level of the patient's anxiety compared to the control condition. The combination of scent and music was not effective in reducing anxiety.

Table 1. *Cont.*

No.	Reference	Sensory				Subject	Tool	Experimental Design	Environmental Procedure	Measures	Main Conclusions
		Visual	Acoustical	Thermal	Smell						
18	Morrin and Chebat, 2005 [76]		✓		✓	774 (Range: >18). Recruited from the mall intercept procedure. Montreal, Canada	Questionnaire Acoustical: mall speaker Scent: diffuser	Acoustical: (music present vs. absent) slow tempo music Olfactory: (scent present vs. absent) citrus scents	In suburban shopping malls: Pretest of scent and sound Poster for participants' recruitment Questionnaire after shopping	Perceived quality of products The mood was measured with the first 2 dimensions of Mehrabian and Russell's (1974) PAD scale The environmental quality of the mall was assessed based on Fisher's (1974) scale	Atmospheric cues such as music and scent were more effective at enhancing consumer response when they were congruent with individuals' affectively or cognitively oriented shopping styles.
19	Ba and Kang, 2019 [78]		✓		✓	168 (54.8% females, M_{age} = 22 (SD = 2.6; min = 18; max = 27)). Recruited from universities via the Internet and by personal contacts. Audition and olfaction screening; no mental illness; and not pregnant. China	Questionnaire Acoustical: loudspeaker Scent: essential oils and perfume	Acoustical: 3 types (birds, conversation, and traffic) Olfactory: 4 types (lilac, osmanthus, coffee, and bread)	In a sound insulation chamber: Pretest of scent and sound Sound evaluation segment; 9 audios, 40 s each Odor evaluation segment; 12 odors, 40 s each Overall evaluation segment 40 s each 5 min ventilation in between for odor and overall evaluation segments	Acoustical comfort, sound preference, sound familiarity, and subjective loudness Olfactory comfort, odor familiarity, and subjective intensity Overall comfort	In the presence of birdsong and low-volume sound, overall comfort and congruency are unaffected by odor. For other combinations of sound and odor, with the increase in concentration, the overall evaluation gradually improves. A positive sensory stimulus can improve the evaluation of perception through other senses, while a negative sensory stimulus has the opposite effect. There is a masking effect between audition and olfaction.
20	Chang et al., 2023 [80]			✓	✓	81 (23 males, 58 females; age range: 18–26). Recruited online in advance; olfaction screening; neither non-smokers nor drug users; have lived locally for more than one year; all volunteers' clothing insulation ranged from 0.16 to 0.72 clo; Xi'an, China	Biometric measures Questionnaire Thermal: outdoor environment Scent: essential oils nebulizer	Thermal: 3 typical spaces The open square (OS) is paved with granite, devoid of vegetation, and unshaded by buildings. The tree-shaded space (TS) is shaded by beeches and its surface is composed of cement pavement and a grass lawn The landscape pavilion (LP) is surrounded by vegetation Olfactory: Lavandula officinalis, Rosa rugosa, and Mentha canadensis	All 3 fragrance stimuli were applied in each measured site for 3 experimental combinations 15 min of adapting to the ambient temperature and completing the first questionnaire Experience fragrance stimuli and were asked to complete the second questionnaire from 3 to 10 min of scent exposure 15 min later, volunteers completed the third questionnaire Repeated the described process until they had visited all sites and experienced all fragrance stimuli	Electroencephalogram (EEG) measures Positive and Negative Affect Schedule (PANAS) Thermal perception vote (thermal sensation vote (TSV) and thermal comfort vote (TCV)) Fragrance perception vote (fragrance sensation vote (FSV), fragrance pleasantness vote (FPV), and fragrance comfort vote (FCV)) Physiological equivalent temperature (PET) and mean TSV (MTSV)	Improving olfactory comfort can partially relieve thermal discomfort caused by high Ta in summer. Fragrance comfort and fragrance pleasure were improved with an increase in thermal comfort. Exposed to R. rugosa and L. officinalis, thermal discomfort may produce a "revenge effect" on fragrance comfort, resulting in fragrance discomfort. Fragrance stimuli increased the beta-band when 30.80 °C ≤ PET < 44.53 °C. When 44.53 °C ≤ PET < 58.27 °C, the alpha-band decreased significantly due to fragrance stimuli. Under different PETs, the relative theta-band in all cerebral cortex zones changed significantly, and the wave band was most significantly influenced by olfactory stimuli.
21	Yang and Moon, 2019 [73]	✓	✓	✓		60 (30 males, 30 females). Recruited from university; vision and hearing screening; they were asked to wear a clothing ensemble of nearly 0.75 clo according to the ASHRAE Standard 55–2004. South Korea	Biometric measures Visual: fluorescent lighting Acoustical: loudspeaker Thermal: room temperature	Visual: 3 illuminance level Acoustical: 4 types of sound (babble, fan, music, and water) with 4 sound levels Thermal: 20, 25, and 30 °C	3 thermal sessions: 30 min of adaptation period 15 min of response [for each sound, 25 s stimulus, 50 s response] 20 min of wash-out period for each illuminance level (3 illuminance levels)	Acoustic comfort Thermal comfort Visual comfort Indoor environmental comfort [using an 11-point numeric scale recommended by ISO 15666] [88]	The effect of acoustic factors was the greatest on indoor environmental comfort, followed by room temperature and illuminance.

Table 1. *Cont.*

No.	Reference	Sensory: Visual	Acoustical	Thermal	Smell	Subject	Tool	Experimental Design	Environmental Procedure	Measures	Main Conclusions
22	Du et al., 2023 [83]	✓	✓	✓		458 (Age range: >60) Recruited from the field; vision and hearing screening; Xi'an, China	Questionnaire Visual: outdoor view Acoustical: loudspeaker Thermal: outdoor temperature	Visual: Visible green index (VGI) Acoustical: 5 common types of stimulating sound (conversation, birdsong, traffic sound, dance music, and traditional opera) 3 ranges LAeq, low (40–45 dBA), medium (50–55 dBA), and high LAeq (60–65 dBA) Thermal: 4 selected spaces: square adjacent to water (WS); tree-shaded square (TS); landscape pavilion (LP); open square (OS); 9:00–11:30, 12:30–15:00, 15:30–18:00	15 min of random type of sound stimulation and fill out the questionnaire after Transfer to the next field and repeat 5 times until listening to all types of sounds	Meteorological parameters, illumination intensity (LUX) Sound types (STP) A weighted equivalent continuous sound pressure level (LAeq) Sky view factor (SVF) Visible green index (VGI) Thermal sensation vote (TSV), thermal comfort vote (TCV), acoustic sensation vote (ASV), acoustic comfort vote (ACV), sunlight sensation vote (SSV), visual comfort vote (VCV), and overall comfort vote (OCV)	When PET was above 43.80 °C, the elderly felt thermally uncomfortable. Older adults perceived traffic sound as acoustically uncomfortable when LAeq was higher than 66.1 dBA. A higher VGI decreased the sensitivity of respondents to LUX. TSV and TCV were susceptible to the acoustic and visual environments. The influence of the visual environment and PET on ASV and ACV were not significant. There was a significant correlation between PET and SSV. Acoustic and thermal comfort had a one-vote veto tendency relative to overall comfort but no absolute veto power. Thermal comfort was the most important factor affecting overall comfort in summer while acoustic comfort was the most important in spring. A binary logistic regression to predict the overall comfort of elderly adults had 84.7% accuracy, indicating a good performance.
23	Sona et al., 2019 [84]	✓	✓		✓	122 (58 males, 64 females; M_{age} = 22.69, SD = 2.23). Recruited from German students. No allergies to the scents used. Germany	Biometric measures Visual: LED screen Acoustical: LED screens with speakers Scent: dispenser	Visual: an artificial window, Acoustical: consisting of 3 high-resolution LED screens with speakers Olfactory: a scent composed of rosewood, geranium, ylang-ylang, olibanum (frankincense), and hyssop in the natural outdoor condition and composition of rosewood and cardamom in the built indoor condition	The 5 conditions are as follows: (1) Control, (2) Nature, (3) Lounge, (4) Scented nature, and (5) Scented lounge 50 min depletion phase 15-min restoration phase and fill in the questionnaire Post-restoration phase	Perception of the space Pleasantness of window view, sound, and odor Perceived Restorativeness Scale Personal resources Fatigue, mood, and arousal	Analyses showed that the subarachnoid hemorrhage early brain edema score (SEBEs) simulating either a natural or a lounge environment was perceived as more pleasant and restorative (fascination/being away) than a standard break room, which in turn facilitated the recovery of personal resources (mood, fatigue, and arousal).

24	Marcus et al., 2019 [85]	✓	✓		✓	154
						(Age range: 18-50;
						city (n = 50, 28 females, 22 males, M_{age}: 27);
						park (n = 52, 26 females, 26 males, M_{age}: 28);
						forest (n = 52, 28 females, 24 males; M_{age}: 27)).
						Recruited from Stockholm (about 1.5 million inhabitants) where they are presumably exposed to a higher degree of the city.
						Vision, hearing, and olfaction screening; inclusion criteria comprised self-declared health, not pregnant, and not using prescription medication.
						Stockholm, Sweden
							Biometric measures
							Questionnaire
							Visual:
							2D 360° virtual reality photo
							VR mask (Oculus Rift)
							Acoustical:
							headphones
							Scent:
							a custom-built nine-channel air-dilution olfactometer
								Visual:
								a densely built-up urban area, a park, or a forest
								Acoustical:
								city noise: traffic; park noise: one bird; forest noise: nine bird species and the sound of a slight breeze)
								Olfactory:
								city odors: diesel, tar, and gunpowder; park odors: grass; forest odors: 2 evergreen species and mushroom
									30 seconds baseline measurement
									150 s stress induction period (shock at 40, 50, 70, 100, and 150 s).
									180 s recovery period
										The average perceived pleasantness
										Physiological stress test (SCL–measures)
										Stress Sensitivity Scale
										SCL (skin conductance level)
											The park and forest, but not the urban area, provided significant stress reduction.
											High pleasantness ratings of the environment were linked to low physiological stress responses for olfactory and auditory but not for visual stimuli.
											Olfactory stimuli may be better at facilitating stress reduction than visual stimuli.

Table 1. *Cont.*

No.	Reference	Sensory				Subject	Tool	Experimental Design	Environmental Procedure	Measures	Main Conclusions
		Visual	Acoustical	Thermal	Smell						
25	Qi et al., 2022 [86]	✓	✓		✓	308 (13% male; Mage: 22.92 (SD = 2.20; Range: 18–31). Recruited from a single college campus. Vision, hearing, and olfaction screening; Shaanxi, China	Biometric measures Questionnaire Visual: 360° virtual reality photo Acoustical: mini wireless speakers Scent: odor sensor	Birdsong only vs. birdsong + photo (4 types)/odor (4 types) vs. birdsong + photo + congruent odor (4 types) Visual: a white wall as the control; short-cut lawn; rose garden; osmanthus tree garden; pine forest Acoustical: Birdsong: downloaded from the open sources on the Internet Streptopelia decaocto (500–800 Hz); riolus chinensis (1 k-3 kHz); Passer montanus (3 k-4 k Hz); Chloris sinica (4 k-6 k Hz); Garrulax canorus (2 k-6 k Hz) Olfactory: leaves from the lawn; flowers of rose bushes; flowers of osmanthus trees; leaves (pine needles) of pine trees	5 min introduction and questionnaire (part 1) 5 min relaxation 1 min baseline 2 min stimulation Questionnaire (part 2-4) 5 min rewards distribution Ventilation	ST (skin temperature), SCL (skin conductance level), and EEG (electroencephalogram) STAI-S (the state version of the State-Trait Anxiety Inventory) Semantic differentials (SDs) survey concerning the overall quality evaluation of the environment (overall attraction, overall harmony, and overall preference)	Integrating visual stimuli of birdsong improved physiological restoration and the overall perceived quality evaluation but held no psychological effect. Introducing olfactory stimuli of birdsong had an adverse restoration physiologically and no significant effect on psychological restoration and the overall preference but enhanced the perceived overall feelings of attraction to the landscape and a sense of overall harmony. Introducing a combination of visual–olfactory stimuli led to increased physiological restoration (only for β-EEG) and overall perceived quality evaluation but also had no significant effect psychologically.
26	Zhong et al., 2022 [87]	✓	✓		✓	172 (78 males, 94 females; Mage: 21). Recruited from the Architecture Department of Chongqing University. Normal hearing and a basic knowledge of soundscapes and landscapes. Chongqing, China	Questionnaire Visual/Acoustical/Scent: outdoor environment	Sense walking Visual/Acoustical/Olfactory: seven waterfront spaces in mountainous cities (WSMCs) in Chongqing were randomly selected as the study areas, namely, Jiangbeizui (JB), Shacixiang (SC), Chaotianmen Square (CT), CBD Riverside Park (CB), Liziba Park (LZ), Jiulongpo Park (JL), and Nanbin Park (NB)	A researcher walking alone or with one or more participants Each participant spent 5 min at each of the walking points to evaluate the soundscape quality and fill out the questionnaire	List all of the sound sources they noticed (referred to the suggestions in ISO/TS 12913-2: 2018 [89]); soundscape comfort degree (SCD) Visual environment comfort degree (VECD); visual environment natural degree (VEND); visual environment diversity degree (VEDD) Smell environment comfort degree (SECD); name the main odors II	In terms of visual elements, the proportions of paved ground, pedestrians, and buildings had negative effects on the soundscape, while those of the sky, water, and natural terrain had positive effects. High visual and smell environment quality can enhance soundscape evaluations, although the smell environment had a greater impact on the SCD than the visual environment in WSMCs.

4.1. Study Type and Subject

Due to differences in research questions, sample size, experimental procedures, and people's behavioral responses, different researchers have adopted different research designs. According to statistics, 20 studies were experimental studies, and 6 studies were field studies; the majority of papers conducted short-term laboratory studies in a controlled environment, where environmental factors other than the target sensory elements were under control. The specific study questions and study sample size are shown in Table 2.

Table 2. Study type and population analysis table.

Project	N(Subject) and M:F and Mean Age and Type				Reference	Remark
Visual Acoustical	20–40	32(16:16)	30	Ordinary people	Park et al., 2020 [50]	
Visual Acoustical	20–40	20(15:5)	27.2	Ordinary people	Hong and Jeon, 2013 [51]	
Visual Acoustical	20–40	40(22:27)	24.1	Students	Jahncke et al., 2015 [52]	
Visual Acoustical	20–40	30; 15; 30 (1:1)	25	Students	Ma and Shu, 2018 [53]	
Visual Acoustical	20–40	31(16:15)	27.9	Ordinary people	Abdalrahman and Galbrun, 2020 [62]	
Visual Acoustical	20–40	38(19:19)	36.3	Students and staffs	Galbrun and Calarco, 2014 [61]	
Visual Acoustical	20–40	28(14:14)	30.1	Ordinary people	Liu et al., 2023 [55]	
Visual Acoustical	20–40	37(18:19)	/	Students	Aristizabal et al., 2021 [42]	
Visual Acoustical	68(40:28)		30–40	Ordinary people	Sun et al., 2018 [54]	Increase the statistical power for subgroup analysis in terms of gender, age, and education
Visual Thermal	Study 1: 19 Study 2: 16 All females		22.2	Students and ordinary people	Kulve et al., 2018 [64]	All participants went through three room-temperature settings
Visual Thermal	84(42:42)		18–30	Students	Chinazzo et al., 2019 [65]	Each participant only experienced one out of three temperature levels
Visual Thermal	86(43:43)		/	Students	Ko et al., 2020 [67]	Compare the window and windowless environments at the same room temperature
Visual Olfactory	21 All females		21.1	Students	Song et al., 2019 [68]	
Visual Olfactory	48(24:24)		22.7	Students	Li et al., 2024 [69]	
Acoustical Thermal	54(25:29)		22	Students	Yang and Moon, 2019 [74]	Participants underwent three experiments
Acoustical Olfactory	247(1:3)		less 20> 65%	Ordinary people	Mattila and Wirtz, 2001 [75]	Field studies can easily recruit participants, as anyone stepping into the test area could be a candidate
Acoustical Olfactory	117(28:89)		47.9	Patients	Fenko and Loock, 2014 [77]	Field studies can easily recruit participants, as anyone stepping into the test area could be a candidate
Acoustical Olfactory	774(/)		/	Shoppers	Morrin and Chebat, 2005 [76]	Field studies can easily recruit participants, as anyone stepping into the test area could be a candidate
Acoustical Olfactory	168(76:92)		22	Students and ordinary people	Ba and Kang, 2019 [78]	Laboratory study
Thermal Olfactory	81(23:58)		18-26	Ordinary people	Chang et al., 2023 [80]	Field study
Visual Acoustical Thermal	60(30:30)		/	Students	Yang and Moon, 2019 [73]	Participants underwent three experiments
Visual Acoustical Thermal	458(/)		>60	The elderly	Du et al., 2023 [83]	Field study
Visual Acoustical Olfactory	122(58:64)		22.69	Students	Sona et al., 2019 [84]	Each participant was exposed to a multisensory environment
Visual Acoustical Olfactory	154(82:72)		27/28	Ordinary people	Marcus et al., 2019 [85]	Each participant was exposed to a multisensory environment
Visual Acoustical Olfactory	308(40:268)		22.92	Students	Qi et al., 2022 [86]	Each participant was exposed to a multisensory environment
Visual Acoustical Olfactory	172(78:94)		21	Students	Zhong et al., 2022 [87]	Field study (sense walking)

Due to limited resources and the high costs associated with maintaining a controlled environment, laboratory studies typically involve fewer than 100 participants (except for acoustical–olfactory experiments). However, as the complexity of multisensory experiments and the number of experimental groups increases, so does the number of participants.

When the number of senses increased to three or more, between-subject experiments required at least a hundred people, and only the intended within-subject experiments used 60 people as a sample.

When we look at the form of recruitment, we find that more than half of the studies were recruited from universities (57%), and a few others were recruited from special approaches and organizations such as online, on-site, and in hospitals (Figure 3). Similarly, participants recruited from the university were overwhelmingly students and mostly under 25 years old, with only one study explicitly stating that it recruited staff from the university. Some studies recruited participants from specific groups, such as the elderly over 60 years of age, patients in hospitals, and shoppers who came to the mall (Figures 4 and 5).

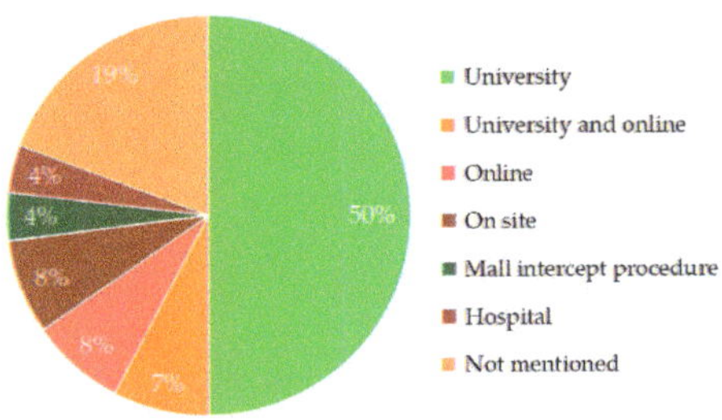

Figure 3. Recruitment method.

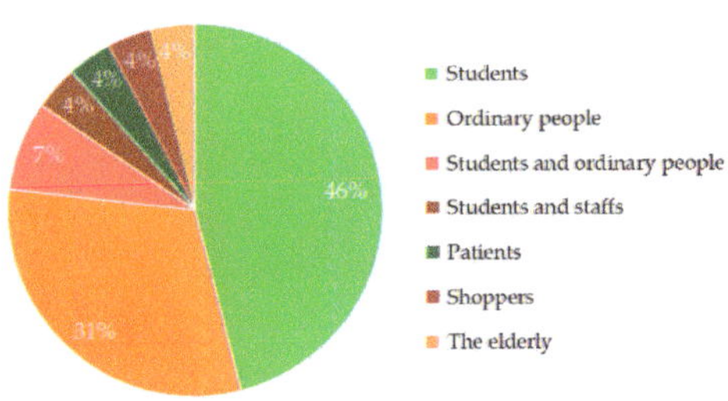

Figure 4. Types of participants.

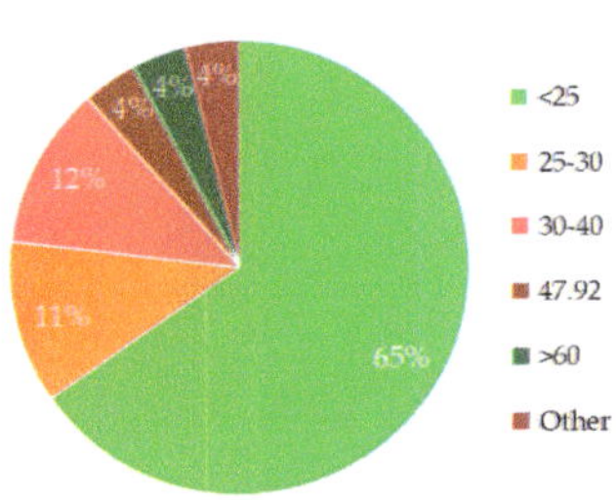

Figure 5. Average age distribution[5].

Due to the consideration of control variables, the vast majority of experiments restrict the characteristics of the population very strictly, such as age range, sex ratio, height and weight, etc. Of the 24 experiments that explicitly stated the sex ratio, 50% (12) studied the ratio of men to women as basically 1:1, and there were even two studies that only studied the female group (Table 2), thus only two studies analyzed the heterogeneity of the population of different ages. Galbrun and Calarco have statistically analyzed cultural groups, such as "Middle Eastern", "Asian", and "African"; however, cultural differences were not statistically analyzed due to small sample sizes for each cultural group.

In addition to screening for age and gender, studies similarly screened participants for sensory abilities, with 75% (15) of the 20 experiments involving screening subjects for hearing ability, either through self-reporting or testing with an audiometer. Only about half of the studies involving the senses of vision and smell screened for their visual and olfactory abilities. Notably, for the study on thermal sensation, Chang et al., 2023 [80] chose residents who had been in the area for more than a year to ensure that they had similar adaptations to the local climate; Chang et al., 2023 [80] and Yang and Moon, 2019 [73,74] achieved similarity in heat perception by controlling for the type of clothing worn by the participants. In addition, mental problems, smoking and drinking, health level, BMI index, pregnancy, medication use, etc., also played a very important role in the inclusion and exclusion criteria of the study. Furthermore, Galbrun and Calarco, 2014 [61] and Hong and Jeon, 2013 [51] conducted a consistency test on the participants to ensure the accuracy of the experimental results.

According to the countries where the experiments were conducted, a total of 11 countries from three continents were involved in the 26 papers, with Asia accounting for 46% (12 papers), Europe for 38% (10 papers), and North America for 15% (4 papers); there was a lack of research on countries in South America, Oceania, and Africa (Figure 6).

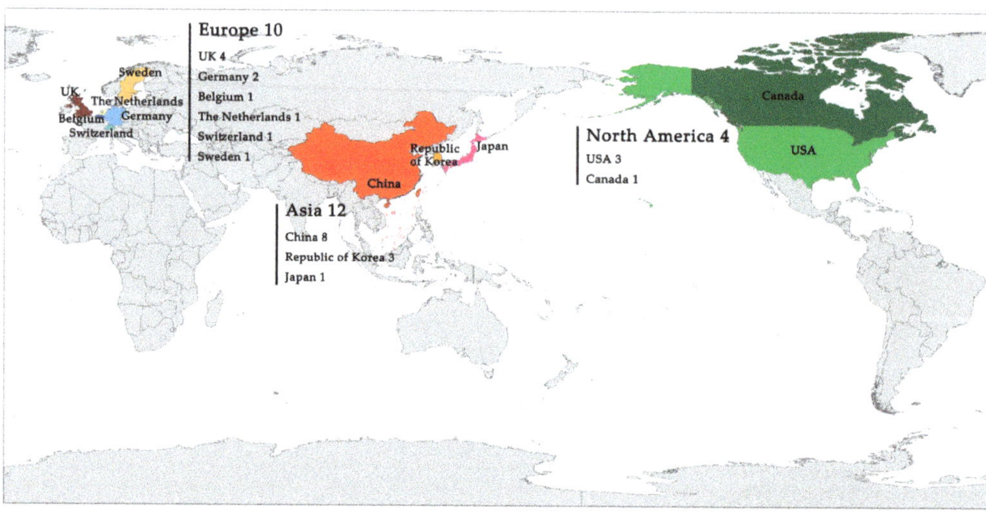

Figure 6. Map of study sites.

4.2. Environmental Settings

Multiple biophilic design strategies could be applied to introduce nature indoors and create sensory experiences in experimental settings. Among the selected studies, researchers adopted different biophilic elements and their combinations, which will be introduced in the following section (Table 3).

Table 3. Environmental settings analysis table.

Types of Elements	Content	Reference	Environmental Settings	
Visual	Nature scenery or natural elements	Park et al., 2020 [50]; Hong and Jeon, 2013 [51]; Jahncke et al., 2015 [52]; Ma and Shu, 2018 [53]; Sun et al., 2018 [54]; Abdalrahman and Galbrun, 2020 [62]; Galbrun and Calarco, 2014 [61]; Ko et al., 2020 [67]; Sona et al., 2019 [84]; Du et al., 2023 [83]; Zhong et al., 2022 [87]		
	Pictures of landscapes or urban nature	Park et al., 2020 [50]; Hong and Jeon, 2013 [51]; Jahncke et al., 2015 [52]; Sun et al., 2018 [54]; Liu et al., 2023 [55]; Aristizabal et al., 2021 [42]; Song et al., 2019 [68]; Marcus et al., 2019 [85]; Qi et al., 2022 [86]		
	Photomontages with different natural features	Abdalrahman and Galbrun, 2020 [62]; Galbrun and Calarco, 2014 [61]		
	Video sequence of a park in the natural outdoor condition	Sona et al., 2019 [84]		
	Greenery existing in offices, such as plants and window view	Ma and Shu, 2018 [53]; Ko et al., 2020 [67]; Li et al., 2024 [69]		
Acoustical	Recording sound live	Park et al., 2020 [50]; Jahncke et al., 2015 [52]; Sun et al., 2018 [54]	Environmental sound: all sounds present in the target environment were included	
		Hong and Jeon, 2013 [51]; Ma and Shu, 2018 [53]; Abdalrahman and Galbrun, 2020 [62]; Galbrun and Calarco, 2014 [61]; Sona et al., 2019 [84]; Du et al., 2023 [83]	Single-resource sound: only one type of sound resource exists when recording the sound, such as water sound, bird sound, and the wind sighing in the trees	
	Network download sound	Marcus et al., 2019 [85]; Qi et al., 2022 [86]		
	Rhythmic music	Mattila and Wirtz, 2001 [75]; Fenko and Loock, 2014 [77]		
Olfactory	Flora scent	Song et al., 2019 [68]	Hinoki cypress leaf oil	
		Mattila and Wirtz, 2001 [75]; Fenko and Loock, 2014 [77]	Lavender	
		Ba and Kang, 2019 [78]	Lilac and Osmanthus	
		Li et al., 2024 [69]	Coriander	
		Chang et al., 2023 [80]	Lavandula officinalis, Rosa rugosa, and Mentha canadensis	
		Qi et al., 2022 [86]	Leaves from the lawn; flowers of rose bushes; flowers of osmanthus trees; leaves (pine needles) of pine trees	

Table 3. *Cont.*

Types of Elements	Content	Reference	Environmental Settings	
		Marcus et al., 2019 [85]	Grass, European silver fir, mushroom from Octanol	
	Wood and herb scent	Sona et al., 2019 [84]	A scent composed of rosewood, geranium, ylang-ylang, olibanum, and hyssop	
	Fruit scents	Mattila and Wirtz, 2001 [75]	Grapefruit	
		Morrin and Chebat, 2005 [76]	Citrus	
	Food scent	Ba and Kang, 2019 [78]	Coffee and bread	
	Urban scent	Zhong et al., 2022 [87]	Natural odors, emission odors, food odors, building material odors, and human odors	
		Marcus et al., 2019 [85]	Diesel, tar, and gunpowder	

For visual exposure, a large majority of studies used screens or VR to display the selected image. However, Park et al. used virtual reality to show the recorded urban video [50]. Yin et al. used virtual reality technology to watch immersive 360-degree field-of-view videos to create a more immersive experience [90]. Instead of using pictures or pre-recorded videos, Yin et al. stimulated three-dimensional virtual offices in VR by using Rhino5 software in advance and rendered in real-time during experiments by using Unity software (version 2017.1.0f3) [9,10]. For acoustical exposure, researchers played sound either through headphones or screen speakers in lab studies and used portable speakers to play the stimulus in field studies. For thermal-related exposure, studies were conducted in lab settings by changing room temperature directly. Experimental sessions were separated based on different temperatures given that it would take a long time to switch between temperatures, and also a long time is needed for participants to adapt. Only one study changed the temperature by directing subjects to three typical spaces [80,83]. For olfactory exposure, the most common way to disperse the scent is using a diffuser, but essential oil and perfume can also be used to create a scented environment.

4.3. Measures

We categorized all measures into five domains: psychological measures, cognitive function, biometrics, comfort, and human behavior (Table 4).

Psychometric methods have their earliest origins and are widely used, and with the development of technology, some studies also incorporated biometric measures along with psychological indicators, because physiological measures can help the researcher understand human responses objectively in real time. Studies in recent years have shown that more and more researchers are using physiological indicator measures, and their methods are more diverse. Cognitive tests were also incorporated into experiments to measure participants' performance. Research on comfort has grown rapidly, especially in the last few years and especially with respect to thermal comfort. Along with physiological and psychological measures, researchers also recorded human social behavior in field studies.

Table 4. Measurement method analysis table.

Measure	Content	Reference	Remark
Psychological measures	Sensation, acceptability, pleasure, familiarity, and subjective intensity	Park et al., 2020 [50]; Hong and Jeon, 2013 [51]; Sun et al., 2018 [54]; Abdalrahman and Galbrun, 2020 [62]; Galbrun and Calarco, 2014 [61]; Aristizabal et al., 2021 [42]; Kulve et al., 2018 [64]; Chinazzo et al., 2019 [65]; Ko et al., 2020 [67]; Yang and Moon, 2019 [73]; Yang and Moon, 2019 [74]; Mattila and Wirtz, 2001 [75]; Fenko and Loock, 2014 [77]; Morrin and Chebat, 2005 [76]; Ba and Kang, 2019 [78]; Sona et al., 2019 [84]; Chang et al., 2023 [80]; Marcus et al., 2019 [85]; Qi et al., 2022 [86]; Zhong et al., 2022 [87]	Helped to explain participants' attitudes toward providing sensory stimuli
	Preferences	Park et al., 2020 [50]; Hong and Jeon, 2013 [51]; Jahncke et al., 2015 [52]; Sun et al., 2018 [54]; Abdalrahman and Galbrun, 2020 [62]; Galbrun and Calarco, 2014 [61]; Mattila and Wirtz, 2001 [75]; Ba and Kang, 2019 [78]; Chang et al., 2023 [80]	A straightforward way to compare different sensory combinations
	Human restoration	Park et al., 2020 [50]; Jahncke et al., 2015 [52]; Sona et al., 2019 [84]	
	Emotional state	Ma and Shu, 2018 [53]; Ko et al., 2020 [67]; Mattila and Wirtz, 2001 [75]; Morrin and Chebat, 2005 [76]; Sona et al., 2019 [84]; Li et al., 2024 [69]	
	Pressure	Ko et al., 2020 [67]; Aristizabal et al., 2021 [42]; Marcus et al., 2019 [85]	
	Anxiety	Fenko and Loock, 2014 [77]; Qi et al., 2022 [86]	
Cognitive function	Design a target search task	Ma and Shu, 2018 [53]; Aristizabal et al., 2021 [42]	Evaluated participants' task performance
	Modules of Cambridge Brain Sciences	Ko et al., 2020 [67]	Evaluated participants' working memory, concentration, short-term memory, and spatial planning and used self-developed tasks to evaluate creativity performance
	Psychomotor vigilance task (PVT) and spatial working memory span task (SWMS)	Li et al., 2024 [69]	Evaluated cognitive performance
Biometrics	Facial electromyography (fEMG) Respiration rate (RR)	Park et al., 2020 [50]	Recorded several physiological indicators to measure the stress recovery process
	Blood pressure (BP)	Ma and Shu, 2018 [53]	
	Heart rate (HR)	Park et al., 2020 [50]; Ma and Shu, 2018 [53]; Aristizabal et al., 2021 [42]; Song et al., 2019 [68]	
	Electrodermal activity (EDA)	Aristizabal et al., 2021 [42]; Li et al., 2024 [69]; Marcus et al., 2019 [85]; Qi et al., 2022 [86]	Skin conductance level (SCL) reflects the activity of the sympathetic nervous system. Used to assess stress levels, with lower levels indicating greater relaxation
	Electrocardiogram (ECG)	Liu et al., 2023 [55]; Li et al., 2024 [69]	
	Electroencephalogram (EEG)	Chang et al., 2023 [80]; Li et al., 2024 [69]; Qi et al., 2022 [86]	Analyzed changes in human mood
	Employ near-infrared time-resolved spectroscopy to measure oxygen-hemoglobin concentrations in the left and right prefrontal cortex of the participants	Song et al., 2019 [68]	Investigated the physiological and psychological relaxation effects
	Body/skin temperature	Kulve et al., 2018 [64]; Chinazzo et al., 2019 [65]; Ko et al., 2020 [67]; Qi et al., 2022 [86]	Analysis of human thermal comfort indicators
	Salivary biochemical indicators: salivary stress marker (cortisol), proinflammatory cytokines, untargeted metabolomics	Li et al., 2024 [69]	

Table 4. *Cont.*

Measure	Content	Reference	Remark
Comfort	Visual comfort	Hong and Jeon, 2013 [51]; Kulve et al., 2018 [64]; Yang and Moon, 2019 [73]; Du et al., 2023 [83]; Zhong et al., 2022 [87]	
	Acoustical comfort	Hong and Jeon, 2013 [51]; Yang and Moon, 2019 [74]; Ba and Kang, 2019 [78]; Yang and Moon, 2019 [73]; Du et al., 2023 [83]; Zhong et al., 2022 [87]	
	Thermal comfort	Kulve et al., 2018 [64]; Ko et al., 2020 [67]; Yang and Moon, 2019 [74]; Chang et al., 2023 [80]; Yang and Moon, 2019 [73]; Du et al., 2023 [83]	
	Olfactory comfort	Ba and Kang, 2019 [78]; Chang et al., 2023 [80]; Zhong et al., 2022 [87]	
	Overall comfort	Chinazzo et al., 2019 [65]; Yang and Moon, 2019 [74]; Ba and Kang, 2019 [78]; Yang and Moon, 2019 [73]; Du et al., 2023 [83]	
Human behavior	Purchasing power	Mattila and Wirtz, 2001 [75]	Evaluated the extent of impulse buying in the experimental environment
	Perceiver quality of products	Morrin and Chebat, 2005 [76]	
	Compared patients' perceived waiting time with the objective waiting time	Fenko and Loock, 2014 [77]	

4.4. Study Procedure

The time for environmental exposure and perception could be different in experiments targeting different senses (Figure 7). For visual and sound perception, humans can process light and sound signals within a second. However, for thermal perception, a human's dermal system needs more time to sense the change in room temperature and make the relevant adaptation. For smell, the chemical composition of the scent needs time to mix with the air and diffuse to smell organs. Due to these differences, studies related to acoustical perception always used a shorter exposure time for each sound stimulus, which is usually less than 15 min and as short as 5 s. However, when considering the recovery and long-term effects, researchers tend to extend the duration of auditory intervention. In a long-term study conducted by Aristizabal et al., a sound file lasting 6 h and 30 min was utilized to replicate an authentic natural environment, which was cyclically repeated [42].

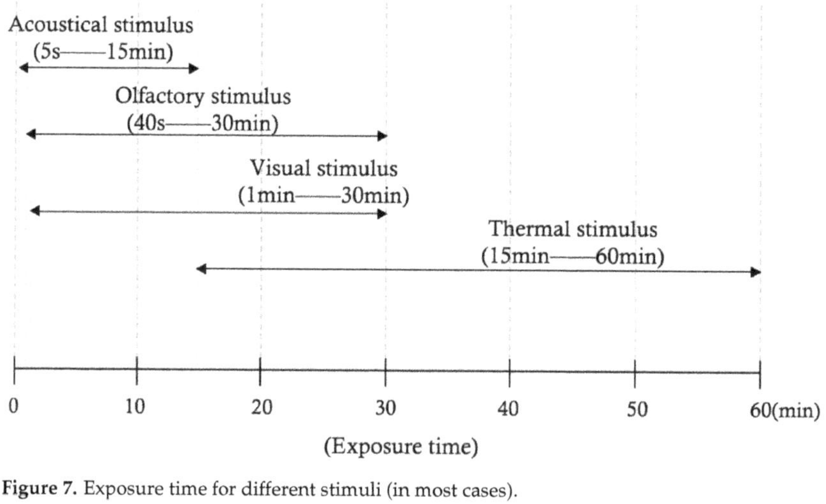

Figure 7. Exposure time for different stimuli (in most cases).

For research assessing visual perception, the exposure time ranged from 1 min to 30 min. The experimental time was always longer than 15 min when changing room temperature was needed. Therefore, if multiple sensory stimuli exist in a single experiment, a longer exposure time is needed for related human responses.

When designing a multisensory study, we not only need to consider the exposure time but also need to consider the carry-over effect of the previous experimental session. For visual and acoustical stimuli, people can quickly switch between different scenes. But when we included temperature as one of the environmental factors, researchers always set aside at least 15 min as adaptation time [64,65,73]. For the lab-based odor study, researchers ventilate the space for 5 min to clear out the residual of the previous scent [73].

5. Gaps and Future Directions

5.1. Beneficial Effects of Sensory Experiences

A noticeable shift has been seen from eliminating undesirable environments to creating beneficial environments. Studies began to look at positive design components advocated by theories of biophilic design, such as incorporating natural view, daylighting, natural sound, air movement, and flora scents indoors. However, we still need more scientific evidence to promote biophilic design. For example, how can we incorporate pleasant scents into the workspaces? How can we use natural sound as a sound masker in offices and public spaces? How can we create dynamic airflow in an office and urban public spaces?

5.2. Interaction between Multisensory Exposure to Nature

A deeper comprehension of interactions among senses would lead to a better understanding, modeling, and prediction of the impacts of sensory elements on people's comfort, perception, health, and well-being. In this review, we examined studies that investigated correlations between different sensory stimuli, including the interactions between visual and acoustical, visual and thermal, acoustical and thermal, and acoustical and olfactory stimuli. However, our findings highlight that studies on the interactions between visual and olfactory, thermal and olfactory, and among three or more senses remain limited. Additionally, tactile interactions with natural elements such as flowers, trees, water, and rocks are acknowledged as potential factors influencing human experience. However, research in this area remains limited. Current studies predominantly focus on the passive tactile perception related to thermal sensations, overlooking the subjective aspects of tactile engagement with these natural elements. This gap highlights the need for further exploration into the active, subjective dimensions of tactile interactions in natural settings.

Beyond these traditional senses, it is important to consider other environmental factors such as humidity and air pressure, which can interact with temperature to influence human sensory experiences. For instance, studies have shown that the combination of temperature and humidity can significantly affect thermal comfort physiological responses [91] and work performance [92]. Moreover, research found that optimal thermal comfort is not static and can depend on several interacting factors, including the direction and speed of airflow [93]. Furthermore, with the advancement of biology, new types of senses—such as emotions (e.g., stress and fear), time perception, and spatial perception—have been identified or defined. Future research should explore these emerging sensory dimensions and their possible connections with traditional sensory modalities. By doing so, we can achieve a more comprehensive understanding of multisensory interactions and their contributions to human well-being, thereby fostering innovation in biophilic design practices.

5.3. Subjects Heterogeneity and Limited Generalization

Existing research has predominantly focused on students or younger populations, with most experiments being conducted in regions like Western Europe, East Asia, and North America. This geographical and demographic concentration limits the generalizability of the findings, thereby restricting their applicability to a broader range of populations. Future studies should expand their focus to include diverse age groups, such as children, middle-aged individuals, and the elderly. Additionally, consideration should also be given to the restorative effects on special populations, such as the chronically ill and pregnant women. By doing so, research can generate more universally applicable conclusions, ultimately fostering a more comprehensive understanding of how biophilic design can contribute to health promotion across various demographics.

5.4. Study Design Methodology for Multisensory Studies

It becomes challenging to design experiments for studying multiple senses. Researchers need to choose between different experimental strategies for different senses. For example, visual studies may need to extend their experimental time to accommodate for the adaption of temperature. For studying the interaction between olfactory and acoustical, researchers need to set aside ventilation time between each pair of sounds and scents.

5.5. The Lasting Impact of Multisensory Experiences

Although most current studies focus on short-term effects (usually within two hours) under various experimental conditions, some studies have begun to explore long-term effects (weeks or months). This is crucial because some benefits of natural exposure are immediate, while others accumulate over extended periods. Therefore, in future studies, it is very necessary to quantify both the immediate and long-term effects of multisensory interaction, particularly exploring the underlying neural mechanisms and psychological processes, as understanding these aspects will be the key to optimizing the biophilic design for human well-being.

5.6. Objective Physiological Measures

The psychological and cognitive effects of environmental stimuli have been studied for more than a century. The processes of data collection heavily rely on subjective self-report factors, which have been criticized for their reliability and precision. Measurement of objective and real-time physiological data via sensors has recently been proposed as a solution to the lack of accuracy of subjective assumptions and has gained some traction, with more than half of the selected biophilic studies already incorporating physiological identification data into their analyses.

To advance our understanding of the complex impacts of biophilic design on human well-being, it is crucial to foster interdisciplinary collaborations. By integrating knowledge and methods from urban design, environmental psychology, public health, and neuroscience, better data interpretation and integration could take place, thereby revealing the intricate relationships between multisensory environmental stimuli and human responses. Such collaborations will not only enhance the accuracy and depth of research findings but also drive innovation and advancements in design practices, ultimately leading to built environments that better support human health and well-being.

6. Conclusions

This review advocates for the accumulation of more scientific evidence to quantitatively evaluate multisensory experiences in biophilic environments and their subsequent influence on health and well-being, thereby underscoring the research's value and future orientation. We therefore comprehensively analyzed empirical studies investigating the interactions between two and three sensory modalities in biophilic design, systematically summarizing the pivotal methods and techniques employed in these experimental studies. These include study types and subject selection, environmental exposure simulations, health benefit assessment indicators, and experimental process design. This analysis not only deepens our understanding of the health benefits associated with biophilic design but also provides methodological insights for future research endeavors.

The existing research shows that multisensory approaches can offer more substantial benefits than those focusing on single sensory modalities. However, current research exhibits limitations in studies involving more than two senses, especially three or more senses, inadequate coverage of different population types and geographical areas, as well as insufficient research on the long-term effects. Therefore, future research should focus on the comprehensive analysis of multisensory interactions. It is essential to consider the restorative effects across various populations and special groups. This inclusive approach will ensure that the benefits of biophilic design are accessible to all, thereby enhancing its impact on public health. At the same time, both short- and long-term impacts on health should be taken into account. Additionally, expanding the use of objective data collection methods and enhancing interdisciplinary collaboration will enable researchers to gain a deeper understanding of this field, ultimately leading to more effective health promotion outcomes.

Author Contributions: J.Y. (Jie Yin): Conceptualization, Methodology, Writing—Review and Editing, Supervision, Project administration, Funding acquisition; H.Z.: Data Curation, Investigation, Writing—Review and Editing; J.Y. (Jing Yuan): Conceptualization, Methodology, Investigation, Data Curation, Writing—Original Draft Preparation. All authors have read and agreed to the published version of the manuscript.

Funding: National Natural Science Foundation of China (No. 52378073); The First Batch of 2024 MOE of PRC Industry-University Collaborative Education Program (Program No. 230805329314629, Kingfar-CES "Human Factors and Ergonomics" Program); Fundamental Research Funds for the Central Universities (No. 22120240378).

Conflicts of Interest: The authors declare no conflicts of interest.

Notes

1 (ASHRAE GUIDELINE 10—Interactions Affecting the Achievement of Acceptable Indoor Environments | Engineering360, n.d.)

2 The checkmarks in the "Sensory" column indicate which sensory elements were examined in each study.

3 Eight participants were excluded before analysis, as they had reported having tinnitus or that they had not heard any sound during the experiment. One participant was excluded for not following the instructions, which left 40 participants for the analysis.

4 In addition, all participants had been living in Minnesota for at least two years by the time the study began and were thus accustomed to the sounds introduced in this study.

5 In this analysis, it is assumed that students whose specific age is not indicated are under 25 years old.

References

1. Wilson, E.O. *Biophilia*; Harvard University Press: Cambridge, MA, USA, 1984.
2. Brown, D.K.; Barton, J.L.; Gladwell, V.F. Viewing Nature Scenes Positively Affects Recovery of Autonomic Function Following Acute-Mental Stress. *Environ. Sci. Technol.* **2013**, *47*, 5562–5569. [CrossRef] [PubMed]
3. Barton, J.; Pretty, J. What is the Best Dose of Nature and Green Exercise for Improving Mental Health? A Multi-Study Analysis. *Environ. Sci. Technol.* **2010**, *44*, 3947–3955. [CrossRef] [PubMed]
4. Park, B.J.; Tsunetsugu, Y.; Kasetani, T.; Kagawa, T.; Miyazaki, Y. The physiological effects of Shinrin-yoku (taking in the forest atmosphere or forest bathing): Evidence from field experiments in 24 forests across Japan. *Environ. Health Prev. Med.* **2009**, *15*, 18–26. [CrossRef] [PubMed]
5. Orsega-Smith, E.; Mowen, A.J.; Payne, L.L.; Godbey, G. The Interaction of Stress and Park Use on Psycho-physiological Health in Older Adults. *J. Leis. Res.* **2004**, *36*, 232–256. [CrossRef]
6. Jahncke, H.; Hygge, S.; Halin, N.; Green, A.M.; Dimberg, K. Open-plan office noise: Cognitive performance and restoration. *J. Environ. Psychol.* **2011**, *31*, 373–382. [CrossRef]
7. Tsunetsugu, Y.; Park, B.-J.; Miyazaki, Y. Trends in research related to "Shinrin-yoku" (taking in the forest atmosphere or forest bathing) in Japan. *Environ. Health Prev. Med.* **2009**, *15*, 27–37. [CrossRef]
8. Kellert, S.R.; Heerwagen, J.; Mador, M. *Biophilic Design: The Theory, Science and Practice of Bringing Buildings to Life*; John Wiley & Sons: Hoboken, NJ, USA, 2011.
9. Yin, J.; Yuan, J.; Arfaei, N.; Catalano, P.J.; Allen, J.G.; Spengler, J.D. Effects of biophilic indoor environment on stress and anxiety recovery: A between-subjects experiment in virtual reality. *Environ. Int.* **2020**, *136*, 105427. [CrossRef]
10. Yin, J.; Arfaei, N.; MacNaughton, P.; Catalano, P.J.; Allen, J.G.; Spengler, J.D. Effects of biophilic interventions in office on stress reaction and cognitive function: A randomized crossover study in virtual reality. *Indoor Air* **2019**, *29*, 1028–1039. [CrossRef]
11. Ulrich, R.S. View through a window may influence recovery from surgery. *Science* **1984**, *224*, 420–421. [CrossRef]
12. Li, D.; Sullivan, W.C. Impact of views to school landscapes on recovery from stress and mental fatigue. *Landsc. Urban Plan.* **2016**, *148*, 149–158. [CrossRef]
13. Felsten, G. Where to take a study break on the college campus: An attention restoration theory perspective. *J. Environ. Psychol.* **2009**, *29*, 160–167. [CrossRef]

14. Aries, M.B.C.; Aarts, M.P.J.; van Hoof, J. Daylight and health: A review of the evidence and consequences for the built environment. *Light. Res. Technol.* **2013**, *47*, 6–27. [CrossRef]
15. Osibona, O.; Solomon, B.D.; Fecht, D. Lighting in the Home and Health: A Systematic Review. *Int. J. Environ. Res. Public Health* **2021**, *18*, 609. [CrossRef]
16. Küller, R.; Lindsten, C. Health and behavior of children in classrooms with and without windows. *J. Environ. Psychol.* **1992**, *12*, 305–317. [CrossRef]
17. Zadeh, R.S.; Shepley, M.M.; Williams, G.; Chung, S.S.E. The Impact of Windows and Daylight on Acute-Care Nurses' Physiological, Psychological, and Behavioral Health. *HERD Health Environ. Res. Des. J.* **2014**, *7*, 35–61. [CrossRef]
18. Benfield, J.A.; Bell, P.A.; Troup, L.J.; Soderstrom, N. Does Anthropogenic Noise in National Parks Impair Memory? *Environ. Behav.* **2010**, *42*, 693–706. [CrossRef]
19. Mathews, K.E.; Canon, L.K. Environmental noise level as a determinant of helping behavior. *J. Pers. Soc. Psychol.* **1975**, *32*, 571–577. [CrossRef]
20. Rautio, N.; Filatova, S.; Lehtiniemi, H.; Miettunen, J. Living environment and its relationship to depressive mood: A systematic review. *Int. J. Soc. Psychiatry* **2018**, *64*, 92–103. [CrossRef]
21. Benfield, J.; Taff, B.; Newman, P.; Smyth, J. Natural sound facilitates mood recovery. *Ecopsychology* **2014**, *6*, 183–188.
22. DeLoach, A.G.; Carter, J.P.; Braasch, J. Tuning the cognitive environment: Sound masking with "natural" sounds in open-plan offices. *J. Acoust. Soc. Am.* **2015**, *137*, 2291. [CrossRef]
23. Ličina, V.F.; Cheung, T.; Zhang, H.; de Dear, R.; Parkinson, T.; Arens, E.; Chun, C.; Schiavon, S.; Luo, M.; Brager, G.; et al. Development of the ASHRAE Global Thermal Comfort Database II. *Build. Environ.* **2018**, *142*, 502–512. [CrossRef]
24. Zhang, H.; Arens, E.; Huizenga, C.; Han, T. Thermal sensation and comfort models for non-uniform and transient environments: Part I: Local sensation of individual body parts. *Build. Environ.* **2010**, *45*, 380–388. [CrossRef]
25. Zhang, H.; Arens, E.; Huizenga, C.; Han, T. Thermal sensation and comfort models for non-uniform and transient environments, part II: Local comfort of individual body parts. *Build. Environ.* **2010**, *45*, 389–398. [CrossRef]
26. Zhang, H.; Arens, E.; Huizenga, C.; Han, T. Thermal sensation and comfort models for non-uniform and transient environments, part III: Whole-body sensation and comfort. *Build. Environ.* **2010**, *45*, 399–410. [CrossRef]
27. Fanger, P.O. Thermal comfort: Analysis and applications in environmental engineering: Fanger, P.O. Danish Technical Press, Copenhagen, Denmark, 1970, 244 pp.: Abstr. in World Textile Abstracts. *Appl. Ergon.* **1972**, *3*, 181. [CrossRef]
28. Huizenga, C.; Hui, Z.; Arens, E. A model of human physiology and comfort for assessing complex thermal environments. *Build. Environ.* **2001**, *36*, 691–699. [CrossRef]
29. Cheng, Y.; Niu, J.; Gao, N. Thermal comfort models: A review and numerical investigation. *Build. Environ.* **2012**, *47*, 13–22. [CrossRef]
30. Zhang, H.; Arens, E.; Zhai, Y. A review of the corrective power of personal comfort systems in non-neutral ambient environments. *Build. Environ.* **2015**, *91*, 15–41. [CrossRef]
31. de Dear, R.J.; Akimoto, T.; Arens, E.A.; Brager, G.; Candido, C.; Cheong, K.W.D.; Li, B.; Nishihara, N.; Sekhar, S.C.; Tanabe, S.-I.; et al. Progress in thermal comfort research over the last twenty years. *Indoor Air* **2013**, *23*, 442–461. [CrossRef]
32. de Dear, R. Revisiting an old hypothesis of human thermal perception: Alliesthesia. *Build. Res. Inf.* **2011**, *39*, 108–117. [CrossRef]
33. Spence, C. Using Ambient Scent to Enhance Well-Being in the Multisensory Built Environment. *Front. Psychol.* **2020**, *11*, 598859. [CrossRef] [PubMed]
34. Jones, A.P. Indoor air quality and health. *Atmos. Environ.* **1999**, *33*, 4535–4564. [CrossRef]
35. Spengler, J.D.; Chen, Q.Y. Indoor air quality factors in designing a healthy building. *Annu. Rev. Energy Environ.* **2000**, *25*, 567–600. [CrossRef]
36. Wargocki, P.; Wyon, D.P.; Sundell, J.; Clausen, G.; Fanger, P.O. The Effects of Outdoor Air Supply Rate in an Office on Perceived Air Quality, Sick Building Syndrome (SBS) Symptoms and Productivity. *Indoor Air* **2000**, *10*, 222–236. [CrossRef]
37. Weber, S.T.; Heuberger, E. The Impact of Natural Odors on Affective States in Humans. *Chem. Senses* **2008**, *33*, 441–447. [CrossRef]
38. Haehner, A.; Maass, H.; Croy, I.; Hummel, T. Influence of room fragrance on attention, anxiety and mood. *Flavour Fragr. J.* **2016**, *32*, 24–28. [CrossRef]
39. Ludvigson, H.W.; Rottman, T.R. Effects of ambient odors of lavender and cloves on cognition, memory, affect and mood. *Chem. Senses* **1989**, *14*, 525–536. [CrossRef]
40. Motomura, N.; Sakurai, A.; Yotsuya, Y. Reduction of Mental Stress with Lavender Odorant. *Percept. Mot. Ski.* **2001**, *93*, 713–718. [CrossRef]
41. Hutmacher, F. Why Is There So Much More Research on Vision Than on Any Other Sensory Modality? *Front. Psychol.* **2019**, *10*, 2246. [CrossRef]
42. Aristizabal, S.; Byun, K.; Porter, P.; Clements, N.; Campanella, C.; Li, L.; Mullan, A.; Ly, S.; Senerat, A.; Nenadic, I.Z.; et al. Biophilic office design: Exploring the impact of a multisensory approach on human well-being. *J. Environ. Psychol.* **2021**, *77*, 101682. [CrossRef]
43. Torresin, S.; Pernigotto, G.; Cappelletti, F.; Gasparella, A. Combined effects of environmental factors on human perception and objective performance: A review of experimental laboratory works. *Indoor Air* **2018**, *28*, 525–538. [CrossRef] [PubMed]

44. Schweiker, M.; Ampatzi, E.; Andargie, M.S.; Andersen, R.K.; Azar, E.; Barthelmes, V.M.; Berger, C.; Bourikas, L.; Carlucci, S.; Chinazzo, G.; et al. Review of multi-domain approaches to indoor environmental perception and behaviour. *Build. Environ.* **2020**, *176*, 106804. [CrossRef]
45. Altomonte, S.; Allen, J.; Bluyssen, P.M.; Brager, G.; Heschong, L.; Loder, A.; Schiavon, S.; Veitch, J.A.; Wang, L.; Wargocki, P. Ten questions concerning well-being in the built environment. *Build. Environ.* **2020**, *180*, 106949. [CrossRef]
46. Arksey, H.; O'Malley, L. Scoping studies: Towards a methodological framework. *Int. J. Soc. Res. Methodol.* **2005**, *8*, 19–32. [CrossRef]
47. Peters, M.D.J.; Godfrey, C.M.; Khalil, H.; McInerney, P.; Parker, D.; Soares, C.B. Guidance for conducting systematic scoping reviews. *JBI Evid.-Based Implement.* **2015**, *13*, 141–146. [CrossRef]
48. Viollon, S.; Lavandier, C. A sound judgment depending on the urban visual setting? *J. Acoust. Soc. Am.* **1999**, *105*, 981. [CrossRef]
49. Viollon, S.; Lavandier, C.; Drake, C. Influence of visual setting on sound ratings in an urban environment. *Appl. Acoust.* **2002**, *63*, 493–511. [CrossRef]
50. Park, S.H.; Lee, P.J.; Jung, T.; Swenson, A. Effects of the aural and visual experience on psycho-physiological recovery in urban and rural environments. *Appl. Acoust.* **2020**, *169*, 107486. [CrossRef]
51. Hong, J.Y.; Jeon, J.Y. Designing sound and visual components for enhancement of urban soundscapes. *J. Acoust. Soc. Am.* **2013**, *134*, 2026–2036. [CrossRef]
52. Jahncke, H.; Naula, S.; Eriksson, K. The effects of auditive and visual settings on perceived restoration likelihood. *Noise Health* **2015**, *17*, 1–10. [CrossRef]
53. Ma, H.; Shu, S. An Experimental Study: The Restorative Effect of Soundscape Elements in a Simulated Open-Plan Office. *Acta Acust. United Acust.* **2018**, *104*, 106–115. [CrossRef]
54. Sun, K.; Sanchez, G.M.E.; De Coensel, B.; Van Renterghem, T.; Talsma, D.; Botteldooren, D. Personal Audiovisual Aptitude Influences the Interaction Between Landscape and Soundscape Appraisal. *Front. Psychol.* **2018**, *9*, 780. [CrossRef] [PubMed]
55. Liu, G.; Zou, J.; Qiao, M.; Zhu, H.; Yang, Y.; Guan, H.; Hu, S. Stress recovery at home: Effects of the indoor visual and auditory stimuli in buildings. *Build. Environ.* **2023**, *244*, 110752. [CrossRef]
56. Watts, G.R.; Pheasant, R.J.; Horoshenkov, K.V.; Ragonesi, L. Measurement and Subjective Assessment of Water Generated Sounds. *Acta Acust. United Acust.* **2009**, *95*, 1032–1039. [CrossRef]
57. Jeon, J.Y.; Lee, P.J.; You, J.; Kang, J. Perceptual assessment of quality of urban soundscapes with combined noise sources and water sounds. *J. Acoust. Soc. Am.* **2010**, *127*, 1357–1366. [CrossRef] [PubMed]
58. Haapakangas, A.; Kankkunen, E.; Hongisto, V.; Virjonen, P.; Oliva, D.; Keskinen, E. Effects of Five Speech Masking Sounds on Performance and Acoustic Satisfaction. Implications for Open-Plan Offices. *Acta Acust. United Acust.* **2011**, *97*, 641–655. [CrossRef]
59. Jeon, J.Y.; Lee, P.J.; You, J.; Kang, J. Acoustical characteristics of water sounds for soundscape enhancement in urban open spaces. *J. Acoust. Soc. Am.* **2012**, *131*, 2101–2109. [CrossRef]
60. Alvarsson, J.J.; Wiens, S.; Nilsson, M.E. Stress Recovery during Exposure to Nature Sound and Environmental Noise. *Int. J. Environ. Res. Public Health* **2010**, *7*, 1036–1046. [CrossRef]
61. Galbrun, L.; Calarco, F.M.A. Audio-visual interaction and perceptual assessment of water features used over road traffic noise. *J. Acoust. Soc. Am.* **2014**, *136*, 2609–2620. [CrossRef]
62. Abdalrahman, Z.; Galbrun, L. Audio-visual preferences, perception, and use of water features in open-plan offices. *J. Acoust. Soc. Am.* **2020**, *147*, 1661–1672. [CrossRef]
63. Hasegawa, Y.; Lau, S.-K. Audiovisual Bimodal and Interactive Effects for Soundscape Design of the Indoor Environments: A Systematic Review. *Sustainability* **2021**, *13*, 339. [CrossRef]
64. Kulve, M.T.; Schlangen, L.; Lichtenbelt, W.v.M. Interactions between the perception of light and temperature. *Indoor Air* **2018**, *28*, 881–891. [CrossRef] [PubMed]
65. Chinazzo, G.; Wienold, J.; Andersen, M. Daylight affects human thermal perception. *Sci. Rep.* **2019**, *9*, 13690. [CrossRef]
66. Haldi, F.; Robinson, D. On the unification of thermal perception and adaptive actions. *Build. Environ.* **2010**, *45*, 2440–2457. [CrossRef]
67. Ko, W.H.; Schiavon, S.; Zhang, H.; Graham, L.T.; Brager, G.; Mauss, I.; Lin, Y.-W. The impact of a view from a window on thermal comfort, emotion, and cognitive performance. *Build. Environ.* **2020**, *175*, 106779. [CrossRef]
68. Song, C.; Ikei, H.; Miyazaki, Y. Physiological effects of forest-related visual, olfactory, and combined stimuli on humans: An additive combined effect. *Urban For. Urban Green.* **2019**, *44*, 126437. [CrossRef]
69. Li, Z.; Zhang, W.; Cui, J.; Wang, L.; Liu, H.; Liu, H. Biophilic environment with visual-olfactory stimuli contributes to psychophysiological restoration and cognitive enhancement. *Build. Environ.* **2024**, *250*, 111202. [CrossRef]
70. Tiller, D.; Wang, L.; Musser, A.; Radik, M. AB-10-017: Combined effects of noise and temperature on human comfort and performance (1128-RP). Faculty Publications in Architectural Engineering. *ASHRAE Trans.* **2010**, *116*. Available online: https://digitalcommons.unl.edu/archengfacpub/40 (accessed on 27 July 2024).
71. Pellerin, N.; Candas, V. Combined effects of temperature and noise on human discomfort. *Physiol. Behav.* **2003**, *78*, 99–106. [CrossRef]
72. Nagano, K.; Horikoshi, T. New comfort index during combined conditions of moderate low ambient temperature and traffic noise. *Energy Build.* **2005**, *37*, 287–294. [CrossRef]

73. Yang, W.; Moon, H.J. Combined effects of acoustic, thermal, and illumination conditions on the comfort of discrete senses and overall indoor environment. *Build. Environ.* **2019**, *148*, 623–633. [CrossRef]
74. Yang, W.; Moon, H.J. Effects of recorded water sounds on intrusive traffic noise perception under three indoor temperatures. *Appl. Acoust.* **2019**, *145*, 234–244. [CrossRef]
75. Mattila, A.S.; Wirtz, J. Congruency of scent and music as a driver of in-store evaluations and behavior. *J. Retail.* **2001**, *77*, 273–289. [CrossRef]
76. Morrin, M.; Chebat, J.-C. Person-Place Congruency. *J. Serv. Res.* **2005**, *8*, 181–191. [CrossRef]
77. Fenko, A.; Loock, C. The Influence of Ambient Scent and Music on Patients' Anxiety in a Waiting Room of a Plastic Surgeon. *HERD: Health Environ. Res. Des. J.* **2014**, *7*, 38–59. [CrossRef]
78. Ba, M.; Kang, J. A laboratory study of the sound-odour interaction in urban environments. *Build. Environ.* **2019**, *147*, 314–326. [CrossRef]
79. Ba, M.; Kang, J. Effect of a fragrant tree on the perception of traffic noise. *Build. Environ.* **2019**, *156*, 147–155. [CrossRef]
80. Chang, J.; Du, M.; Hong, B.; Qu, H.; Chen, H. Effects of thermal-olfactory interactions on emotional changes in urban outdoor environments. *Build. Environ.* **2023**, *232*, 110049. [CrossRef]
81. Krüger, E.L.; Zannin, P.H. Acoustic, thermal and luminous comfort in classrooms. *Build. Environ.* **2004**, *39*, 1055–1063. [CrossRef]
82. Huang, L.; Zhu, Y.; Ouyang, Q.; Cao, B. A study on the effects of thermal, luminous, and acoustic environments on indoor environmental comfort in offices. *Build. Environ.* **2012**, *49*, 304–309. [CrossRef]
83. Du, M.; Hong, B.; Gu, C.; Li, Y.; Wang, Y. Multiple effects of visual-acoustic-thermal perceptions on the overall comfort of elderly adults in residential outdoor environments. *Energy Build.* **2023**, *283*, 112813. [CrossRef]
84. Sona, B.; Dietl, E.; Steidle, A. Recovery in sensory-enriched break environments: Integrating vision, sound and scent into simulated indoor and outdoor environments. *Ergonomics* **2019**, *62*, 521–536. [CrossRef] [PubMed]
85. Hedblom, M.; Gunnarsson, B.; Iravani, B.; Knez, I.; Schaefer, M.; Thorsson, P.; Lundström, J.N. Reduction of physiological stress by urban green space in a multisensory virtual experiment. *Sci. Rep.* **2019**, *9*, 10113. [CrossRef]
86. Qi, Y.; Chen, Q.; Lin, F.; Liu, Q.; Zhang, X.; Guo, J.; Qiu, L.; Gao, T. Comparative study on birdsong and its multi-sensory combinational effects on physio-psychological restoration. *J. Environ. Psychol.* **2022**, *83*, 101879. [CrossRef]
87. Zhong, B.; Xie, H.; Gao, T.; Qiu, L.; Li, H.; Zhang, Z. The Effects of Spatial Characteristics and Visual and Smell Environments on the Soundscape of Waterfront Space in Mountainous Cities. *Forests* **2022**, *14*, 10. [CrossRef]
88. *ISO/TS 15666: 2003*; Acoustics-Assessment of Noise Annoyance by Means of Social and Socio-Acoustic Surveys. International Organisation for Standardization: Geneva, Switzerland, 2003.
89. *ISO/TS 12913-2: 2018*; Acoustics—Soundscape—Part 2: Data Collection and Reporting Requirements. ISO: Geneva, Switzerland, 2018.
90. Yin, J.; Zhu, S.; MacNaughton, P.; Allen, J.G.; Spengler, J.D. Physiological and cognitive performance of exposure to biophilic indoor environment. *Build. Environ.* **2018**, *132*, 255–262. [CrossRef]
91. Nikolopoulou, M.; Steemers, K. Thermal comfort and psychological adaptation as a guide for designing urban spaces. *Energy Build.* **2003**, *35*, 95–101. [CrossRef]
92. Wu, J.; Hou, Z.; Shen, J.; Lian, Z. Quantitative effect on work performance considering interactions among multiple indoor environmental factors. *Build. Environ.* **2020**, *185*, 107286. [CrossRef]
93. Luo, M.; Zhang, H.; Wang, Z.; Arens, E.; Chen, W.; Bauman, F.S.; Raftery, P. Ceiling-fan-integrated air-conditioning: Thermal comfort evaluations. *Build. Cities* **2021**, *2*, 928–951. [CrossRef]

land

MDPI

Article

Exploring the Impact of Visual and Aural Elements in Urban Parks on Human Behavior and Emotional Responses

Tongfei Jin, Jiayi Lu and Yuhan Shao *

College of Architecture and Urban Planning, Tongji University, Shanghai 200092, China;
tongfei_jin@tongji.edu.cn (T.J.)
* Correspondence: shaoyuhan@tongji.edu.cn

Abstract: As cities progress into high-quality developments, the demand for urban parks that enhance residents' well-being and sustainability is increasing. Traditional visual-centric design methods no longer suffice. Given that vision and hearing are the primary sensory pathways through which people perceive their environment, exploring their relationship with landscape experiences offers a novel perspective for optimizing the audiovisual perception quality of urban parks. This study explores the relationship between visual and auditory elements and landscape experiences to optimize urban parks' sensory quality. Using visual perception, soundscape perception, sound source perception, and behavioral vitality, this study evaluates the audiovisual perception quality of a representative wetland park in Chengdu's ring ecological zone. By quantifying relationships between audiovisual characteristics, behavioral vitality, and emotional feedback, several emotional assessment models were constructed. The results show that lawns, pavements, and sound pressure levels significantly impact vitality. A sound pressure level of 77 dB has been identified as a critical threshold in emotional perception models. Consequently, distinct emotional prediction models can be employed to enhance landscape design across various sound pressure level zones. This research provides scientific evidence and flexible strategies for designing urban open spaces that improve landscape experiences based on multisensory perception.

Keywords: visual and aural elements; behavior and emotional responses; perceptions; well-being; sustainable development

1. Introduction

The WHO's "Healthy Cities" initiative emphasizes the importance of urban environments that promote health and well-being [1]. In response, agencies in various countries, such as the U.S. Environmental Protection Agency, National Health Service of the U.K., and the National Health Commission of the People's Republic of China, have published guidelines for designing open spaces that are accessible, safe, and conducive to physical activity and mental relaxation [2–4]. Green and open spaces help reduce stress and anxiety, enhance well-being, and improve the quality of life, directly impacting mental health positively. Parks, as typical natural landscapes in cities, have been widely proven to regulate urban residents' emotions—such as inducing pleasure, alleviating anxiety, and boosting self-confidence—and promote behaviors such as recreation and social interaction. However, due to the numerous intrinsic factors influencing emotional feedback and behavior, many studies have insufficiently addressed the complex interactions between environmental characteristics and people's emotional and behavioral responses.

1.1. The Health Benefits of Parks

Urban parks and green spaces are essential components of sustainable urban environments, offering numerous health benefits to residents. Research consistently demonstrates that exposure to natural environments in urban areas can significantly improve mental and physical health. For instance, studies have shown that green spaces help reduce stress and anxiety, enhance mood, and promote overall well-being [5,6]. Parks provide settings for physical activities such as walking, jogging, and cycling, which are crucial for maintaining physical health and preventing chronic diseases [7,8].

Moreover, the presence of natural elements like trees, water features, and diverse plantings can create restorative environments that facilitate recovery from mental fatigue and stress [9,10]. These environments are particularly effective in urban settings where residents are often exposed to high levels of stress and limited access to natural scenery [11]. The role of parks in fostering social interactions and community engagement is also significant. Well-designed green spaces can act as social hubs where people gather, interact, and build social ties, contributing to social cohesion and community well-being [12,13].

Furthermore, parks contribute to environmental health by improving air quality, reducing urban heat island effects, and providing habitats for urban wildlife, thus enhancing urban biodiversity [14,15]. These ecological benefits, in turn, support human health by creating healthier living environments. Therefore, the integration of parks and green spaces in urban planning is critical for promoting sustainable and healthy urban living [16,17].

1.2. The Effects of Visual or Auditory Elements on Human Behavior and Emotions

Visual elements in urban parks and green spaces significantly influence human behavior and emotional well-being. The presence of natural scenery, such as trees, flowers, and water features, has been shown to enhance mood and reduce stress levels. Research indicates that visual exposure to green environments can lower heart rates and blood pressure, thereby promoting relaxation and recovery from mental fatigue [9,18]. Additionally, the visual presence of water features in urban parks has been linked to increased tranquility and reduced anxiety, further highlighting the importance of incorporating diverse natural elements in urban design [19]. Moreover, the diversity and complexity of visual stimuli in natural environments can enhance cognitive function and creativity, as exposure to varied natural scenes stimulates different areas of the brain [20]. These environments provide a sense of escape from the urban hustle and bustle, contributing to restorative experiences that are essential for mental health [6].

Auditory elements in urban parks and green spaces also play a crucial role in shaping human behavior and emotional responses. Natural sounds, such as birdsong, rustling leaves, and flowing water, have been found to significantly enhance mental well-being and reduce stress. These sounds create a calming and restorative auditory environment that can lower cortisol levels and promote relaxation [21,22]. Research has demonstrated that auditory elements influence the overall perception of environmental quality and satisfaction with urban spaces. For instance, areas with pleasant natural sounds are more likely to be perceived as restorative and are preferred for recreational activities [23,24]. The soundscape can also affect the choice of activities within a park, encouraging behaviors such as leisurely walking, meditation, and social interaction in areas dominated by natural sounds [25]. Conversely, the presence of noise pollution, such as traffic sounds, can detract from these benefits, increasing stress levels and reducing the restorative potential of green spaces [26].

The psychological benefits of visual and auditory elements in urban parks underscore the need for the careful design and maintenance of urban parks. Incorporating a variety of plants, ensuring seasonal changes in scenery, and maintaining cleanliness are essential for maximizing the positive effects on human behavior and emotions [27,28]. Strategies such as creating water features, planting sound-absorbing vegetation, and designing quiet zones away from urban noise sources can enhance the acoustic quality of parks [29]. By creating visually appealing environments and prioritizing natural soundscapes, urban planners can promote healthier and more fulfilling urban lifestyles.

1.3. The Interactive Effects of Visual and Auditory Elements on Human Behavior and Emotions

The interaction between visual and auditory elements in urban green spaces profoundly impacts human behavior and emotions, creating multisensory experiences that contribute to overall well-being. While research has shown that both visual and auditory stimuli individually enhance mental health and promote positive behaviors, understanding their combined effects remains an emerging field. Studies have demonstrated that environments combining visually appealing landscapes with pleasant natural sounds are perceived as more restorative and beneficial than those with a single sensory stimulus [30–32]. This synergistic effect underscores the importance of designing urban spaces that cater to multiple senses to maximize health benefits.

Despite these insights, significant gaps remain in the research. Most existing studies rely heavily on qualitative assessments and correlational analyses, which, although valuable, do not provide a comprehensive understanding of the underlying mechanisms. There is a notable lack of quantitative studies that explore the mathematical relationships between visual and auditory elements and their combined impact on human behavior and emotions [23,33,34]. Without such data, it is challenging to develop predictive models that can inform the design of more effective restorative environments. Furthermore, many studies do not consider the potential moderating factors such as individual differences in sensory processing, cultural backgrounds, and situational contexts that might influence the perceived benefits of multisensory environments [35–37]. Addressing these gaps requires more rigorous experimental designs and advanced analytical techniques.

Considering the research review, this study aims to investigate the integrated effects of visual and aural stimuli from urban nature on human behavior and emotional responses, especially in the context of traffic noise influence. Drawing upon sustainable landscape design, this research will focus on understanding how the visual landscape and auditory environment of urban green spaces influence emotional responses such as satisfaction, pleasantness, and calmness, as well as behaviors observed within these environments.

Using Bailuwan Wetland Park in Chengdu as a case study, this research will utilize spatial analysis techniques, including space syntax and QGis, to quantify visual landscape attributes. A soundwalk and survey will assess perceived aural characteristics, while onsite observations will document prevalent activities. The data will be used to develop a model that explores how visual and auditory stimuli interact to affect behavior and emotions. By integrating findings from both visual and aural perspectives, this study seeks to contribute to the advancement of sustainable landscape design practices that foster positive emotional experiences and encourage beneficial behaviors within urban environments.

2. Materials and Methods

2.1. Research Site

This research was conducted at the Bailuwan Wetland Park, a typical urban green space surrounding the Chengdu Ring Expressway, as depicted in Figure 1. Located in the southeast corner of Chengdu, Sichuan Province, the park is adjacent to an expressway, resulting in significant traffic noise issues and concentrated citizen complaints. The total area of the park is 2 square kilometers, with approximately one-third covered by water. As a national urban wetland park, it primarily serves recreational, sightseeing, and ecological conservation functions. The park features diverse landscape elements, including water bodies, lawns, woodlands, and paved pavements. The current severe noise problems and the public's pursuit of a high-quality living environment pose higher demands for the sustainable development of the area.

Considering the park's construction status, open areas, functional diversity, and environmental sound pressure levels, ten measurement points were selected within the research site to evaluate landscape elements and soundscape satisfaction. The selected points meet the following criteria: (1) Each point has unique visual and acoustic characteristics, distinguishing them from each other. (2) The distance between measurement points is greater than 300 m to eliminate interference from the soundscape of adjacent points. (3) Measurement points cover most visual elements within the research site, offering open views and rich visual elements, suitable for acoustic measurement and visual factor calculations. (4) Each location should have an activity space of no less than 50 m × 50 m, allowing researchers to move freely to gain a more comprehensive acoustic landscape experience, while also not causing excessive differences in the sound environment [38,39].

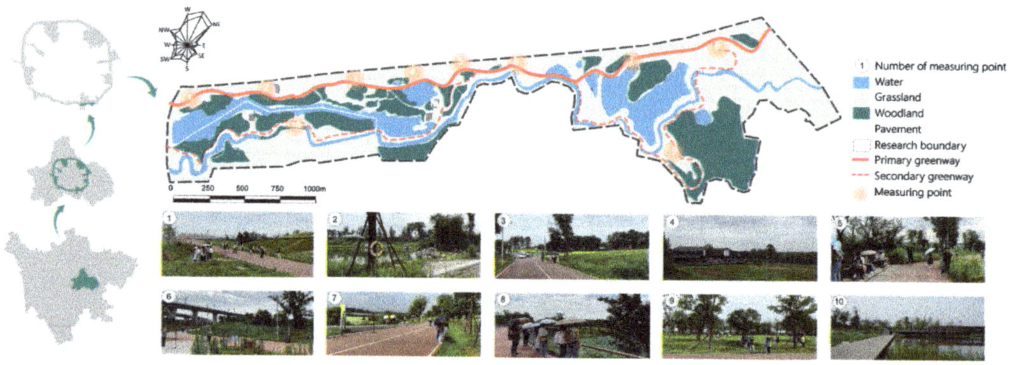

Figure 1. Location map of measurement points in Bailuwan Wetland Park.

Nevertheless, we admit there might be some potential bias regarding site selection. For instance, by focusing on areas with open views and rich visual elements, this study might underrepresent more enclosed or vegetated areas that contribute differently to the overall soundscape. The requirement for a 300 m distance between points could result in the exclusion of smaller but significant park areas, leading to a potential underrepresentation of their soundscape characteristics.

2.2. Measurements

This study constructs an indicator system based on visual and auditory environmental characteristics to explore the impact mechanisms of the Bailuwan Wetland Park landscape on human activities and emotional perception. Using QGIS spatial syntax calculations and field surveys, audiovisual environment data for the ten points were collected. Questionnaire surveys quantified behavioral vitality intensity and emotional perception indicators to explore the relationship between audiovisual environments and human activities and emotions. Finally, multiple regression analysis was used to form a predictive model for optimizing audiovisual environments in ecological parks around urban expressways (Figure 2).

2.2.1. Visual Landscape Composition

Considering the objective existence and subjective perception of visual landscape factors [40], this study selected the proportions of different landscape elements within people's visual fields to represent visual characteristics. Researchers first conducted a site survey of the research area, using a panoramic camera, Gopro, to capture panoramic images of the visual landscape. Through semantic segmentation model (Qgis 3.36.2), the visual landscape elements in the site were ultimately summarized and extracted into four types: woodland (Wo), lawn (L), water bodies (Wa), and pavements (P). Subsequently, geographic information system (QGIS) and spatial syntax methods were used to calculate the composition of the landscape perceived visually by people throughout the entire research area.

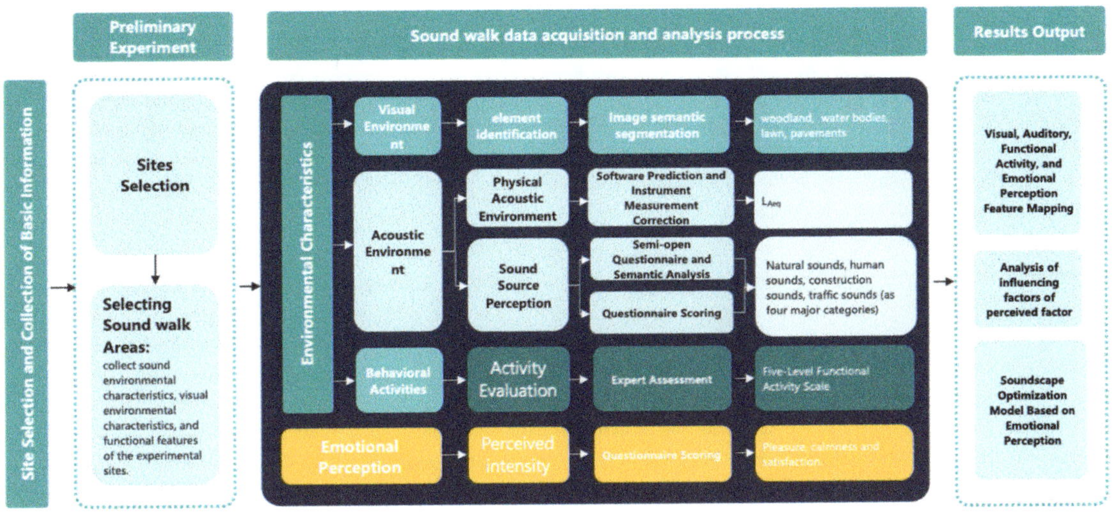

Figure 2. Measurement methods diagram.

2.2.2. Auditory Environmental Characteristics

This study used the equivalent continuous A-weighted sound pressure level (L_{Aeq}) and sound source perception indicators to describe the park's sound environment. L_{Aeq} is a standardized measure for environmental sound pressure levels, expressing decibel values and variations in sound intensity, and has been shown to significantly impact soundscape perception [41]. A multi-channel signal analyzer (AWVA6290L+) recorded and measured the sound environment at each measurement point. Additionally, based on traffic flow data from surrounding expressways and field measurements, a noise map of the site was calculated using the noise prediction model (NPL) [see http://resource.npl.co.uk/acoustics/techguides/crtn/ (accessed on 18 March 2024)]. Sound sources were used to describe which sounds dominate within the study site.

Sound source perception consists of two parts: the type of sound source and the intensity of perception. This study employed a combination of soundwalks and questionnaires to conduct research on sound source perception. A soundwalk is a standardized method widely used in soundscape research to measure auditory perception [42]. This method primarily involves organizing participants to take walks along predetermined routes, during which they focus on experiencing the surrounding soundscape environment and record and evaluate the sounds they hear and their relationship with the surrounding landscape environment at designated research points [43,44]. The sound environment characteristic evaluation part of the questionnaire is divided into two sections. Section 1 is a semi-open questionnaire used to record the types of sound sources at the research points. Section 2 uses a five-point Likert scale (1–5, where 1 represents "not perceived at all" and 5 represents "very strongly perceived") to evaluate the perceived intensity of different types of sound sources. In the pre-experiment, researchers mainly recorded the categories of sound sources to distinguish the differences in the acoustic environment at different research points and to formulate subsequent experimental methods. In the formal experiment, participants were invited to refine the specific types of sound sources. Based on two experiments, the sound sources in Bailuwan Wetland Park can be categorized into four types: traffic sounds (S1), including car horns, rail bicycle rings, motor vehicles, highways, heavy vehicles, and airplanes; construction sounds (S2), including alarms, radios, and loudspeakers; human sounds (S3), including exercising, cleaning, whistling, walking, parent–child activities, talking, laughing, and singing; and natural sounds (S4), including insects, tree murmur, wind, twittering of birds, ripple, fish diving, and dogs barking.

2.2.3. Behavioral Vitality Characteristics

This study used behavioral vitality (V) to quantify the behavioral vitality characteristics of the site. Behavioral vitality intensity refers to the diversity and intensity of activities within different areas, reflecting the impact of the landscape environment on human behavior. These data were collected through field surveys and expert assessments, recording activities lasting over 3 min within the measurement points and rating their diversity using a five-point Likert scale (1–5, where 1 indicates "single activity" and 5 indicates "highly diverse/intense activities").

2.2.4. Emotional Perception Evaluation

Based on the classic Russell complex emotion model [45], the Perceived Affective Quality (PAQ) model, and international soundscape standards [46], previous experiments typically selected pleasure, calmness, and annoyance to describe emotional perceptions in the environment. Satisfaction was also selected to evaluate the overall psychological impact of the coupled audiovisual environment in urban green spaces around expressways. However, considering that calmness and annoyance are antonymous in the Chinese context, this study retains the positive emotional indicators of pleasure (E1), calmness (E2), and satisfaction (E3) as the emotional evaluation indicators for this research. Data were collected through a soundwalk and questionnaires, using a five-point Likert scale (1–5, where 1 indicates "very disagreed" and 5 indicates "very agreed").

2.3. Data Collection

The data collection for this study involved both desk research and field surveys. Initially, desk research was conducted to obtain data on visual environmental characteristics using geographic remote sensing systems. Google satellite images and the Qgis 3.36.2 software platform were employed to acquire remote sensing images of the study site. The proportions and distributions of four types of research elements were measured. Boundaries for each element were delineated using QGIS Maptiler, and the visible areas of landscape elements throughout the entire park were analyzed with Depthmap X, a spatial syntax software, considering buildings and woodland as visual barriers. A grid of 50 m × 50 m, corresponding to the experimental plot space, was applied. Vector circles were generated in QGIS based on the actual positions and ranges of each measurement point, from which the proportions of visual landscape elements were extracted (Figure 3).

Data on auditory environmental characteristics, behavioral vitality characteristics, and psychological perception indicators were primarily collected through field surveys. The surveys were conducted on clear days with no obstructions, rain, snow, or lightning, ensuring good air quality and visibility, and wind speeds below 5 m/s. Observations were made between 8:00 AM and 17:00 PM. To eliminate interference from other factors, 30 participants aged 20–30, with a gender ratio of approximately 1:1, were recruited locally. All participants had good vision and hearing to control for perceptual biases due to individual differences. This study received ethical approval, and informed consent was obtained from all participants. Prior to the survey, participants received necessary training and briefings.

The selection of participants aged 20–30 was driven by this study's focus on controlling for potential perceptual biases related to sensory acuity and cognitive processing, which can vary significantly with age. Young adults generally possess stable and consistent sensory capabilities, which minimizes variability due to age-related sensory decline. This demographic also tends to have a higher level of familiarity with digital interfaces and technology, which was utilized in the data collection process, ensuring that all participants could engage with the survey tools effectively. Additionally, by focusing on a specific age group, this study aims to produce more homogeneous and reliable data, thereby reducing confounding variables and enhancing the internal validity of the findings. Future studies could expand the age range to explore how these factors evolve across the lifespan.

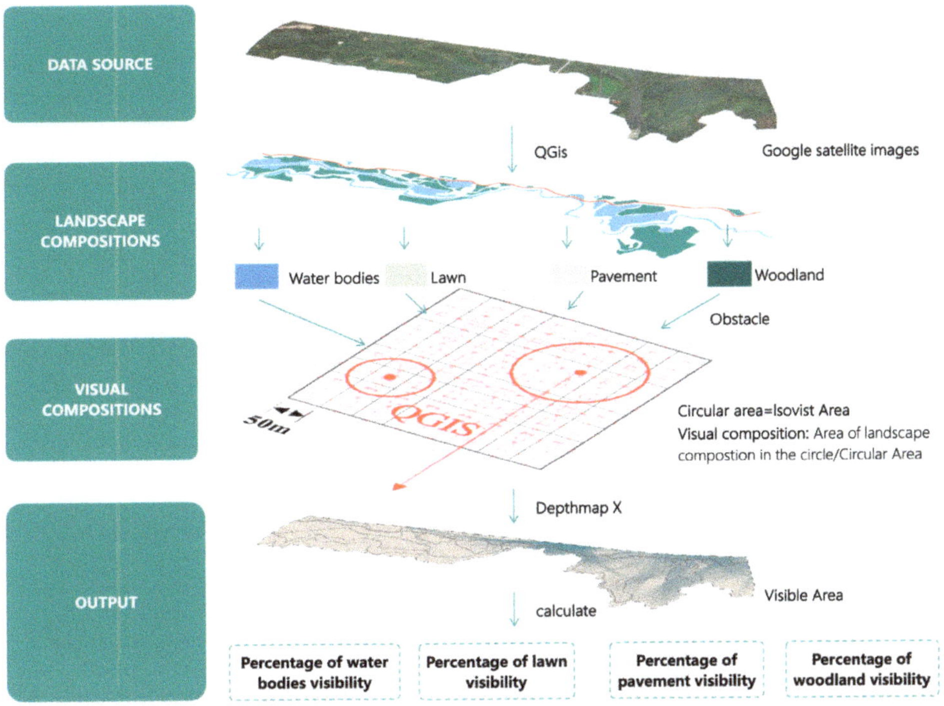

Figure 3. The method of visual landscape calculations of the site.

Once the survey began, 30 participants were divided into five groups. Each group, led by a researcher, followed different paths through the 10 measurement points for soundwalk experiments. Participants freely walked and stayed within the soundwalk areas, observing environmental characteristics and listening to ambient sounds for 5 min [47]. Afterward, they completed and submitted online questionnaires via mobile devices. A total of 250 questionnaires were distributed, and 244 valid responses were obtained, meeting the effect size requirements for visual and soundscape evaluation studies [48].

Simultaneously, researchers measured and evaluated the auditory environment and behavioral vitality characteristics at each point. After calibrating the multi-channel signal analyzer, two amplifiers and microphones equipped with wind shields were placed approximately 1.6 m above ground at the center of each measurement point to record environmental sound pressure levels for 3 min. The equivalent continuous A-weighted sound pressure level (Laeq) was obtained using the multi-channel signal analyzer. Additionally, during the park's opening hours, professional observers recorded the types of functional activities lasting more than 3 min within the measurement points every two hours (10:00, 12:00, 14:00, and 16:00) and rated their diversity.

2.4. Data Analysis and Visualization

There are two main methods to explore the relationship between emotions and audiovisual environmental characteristics. The first method involves correlation analysis to investigate the factors influencing positive and negative emotions, but it cannot show the interactions between these factors. The second method involves constructing relational equations to predict the impact of the overall environment on human activities and emotional feedback based on the quality of audiovisual environments. Based on this, this study employs modeling and map visualization techniques to process the data, providing an intuitive representation of the park's behavioral vitality intensity and emotional responses.

The data analysis in this study can be divided into three main steps. First, we constructed a relational model between audiovisual environmental characteristics and behavioral vitality characteristics, with behavioral vitality characteristics as the dependent variable and objective elements of the audiovisual environment as independent variables, to predict the intensity of functional activities throughout the park. Next, using emotional perception elements (pleasure, calmness, annoyance, and satisfaction) as dependent variables, we analyzed the impact of the proportions of various visual elements, perceived intensities of different sound sources, and behavioral vitality levels, constructing a multiple linear regression model to describe the relationship between landscape environment and emotional perception. This model aims to discover the driving factors that affect emotional changes, providing support for the optimization of the audiovisual environment in ecological parks along urban expressways. Finally, we utilized QGIS to visualize the models, predicting the intensity of functional activities and emotional feedback throughout the park, thereby guiding the sustainable development and health of the park.

3. Results

3.1. Manipulation Checks

The collected questionnaire data were subjected to Shapiro–Wilk and Kolmogorov–Smirnov tests to assess normality, with the results presented in Table 1. A significance level of $\alpha = 0.05$, $p > 0.05$, indicates that the data meet the normal distribution criteria and are suitable for subsequent analysis. Reliability tests were conducted on the collected questionnaires, and the results indicated that the Cronbach's α for the satisfaction evaluation exceeded 0.6, demonstrating the overall validity of the collected data [49]. An independent-samples *t*-test was performed based on gender grouping. The Levene's test for equality of variances yielded values greater than 0.05, indicating the homogeneity of variances. The *t*-test for equality of means also showed significance values greater than 0.05, indicating that gender differences did not affect the evaluation results.

Table 1. Significant (*p*) values of normality test.

	Significant (*p*-Value)	
	Shapiro–Wilk Test	**Kolmogorov–Smirnov Test**
Traffic sounds (S1)	0.290	0.776
Construction sounds (S2)	0.417	0.601
Human sounds (S3)	0.240	0.854
Natural sounds (S4)	0.285	0.762
Behavioral vitality (v)	0.354	0.625
Pleasure (E1)	0.220	0.860
Calmness (E2)	0.178	0.900
Satisfaction (E3)	0.242	0.890

3.2. Descriptive Analysis

The visual environmental characteristics, auditory environmental characteristics, behavioral vitality characteristics, and psychological perception results for the ten measurement points at the study site are shown in Table 2 and Figure 4.

Table 2. Objective environmental and psychological perception data for measurement points.

Point Location		Wo	Wa	L	P	Others	L_{Aeq}	S1	S2	S3	S4	V	E1	E2	E3
NO.1	Mean	34%	1%	52%	13%	0%	75.9	3.77	1.68	2.41	4.68	3	3.14	2.73	3.3
	Std. Dev	-	-	-	-		-	0.73	0.76	0.78	0.47	-	0.69	0.86	0.63
NO.2	Mean	17%	2%	65%	16%	0%	75.9	4.86	1.18	1.29	4.25	3	2.18	1.57	2.43
	Std. Dev	-	-	-	-		-	0.35	0.38	0.52	0.69	-	0.89	0.73	0.98
NO.3	Mean	22%	6%	56%	16%	0%	74.6	4.45	1.17	1.76	4.55	3	2.41	1.93	2.69
	Std. Dev	-	-	-	-		-	0.56	0.38	0.82	0.56	-	0.77	1.14	0.83
NO.4	Mean	29%	14%	45%	12%	0%	71.4	4.42	1.26	2.05	4.42	3	3.05	2.42	3.42
	Std. Dev	-	-	-	-		-	0.67	0.44	0.69	0.59	-	0.69	0.88	0.59
NO.5	Mean	6%	7%	77%	10%	0%	75.1	4.45	1.23	2.55	4.32	0	2.55	1.73	2.82
	Std. Dev	-	-	-	-		-	0.66	0.42	0.58	0.63	-	0.58	0.69	0.65
NO.6	Mean	3%	18%	66%	13%	0%	76.8	4.92	1.71	1.54	4.17	2	1.96	1.42	2.21
	Std. Dev	-	-	-	-		-	0.28	0.93	0.58	0.75	-	0.73	0.64	0.82
NO.7	Mean	18%	17%	54%	11%	0%	79.2	4.41	1.17	1.97	4.29	3	3.07	2.55	3.35
	Std. Dev	-	-	-	-		-	0.67	0.38	0.72	0.61	-	0.94	1.19	0.9
NO.8	Mean	30%	14%	49%	7%	0%	78.5	4.65	1.22	2.04	4.04	3	3.09	2.39	3.21
	Std. Dev	-	-	-	-		-	0.56	0.41	0.81	0.86	-	0.65	1.01	0.63
NO.9	Mean	57%	0%	27%	16%	0%	66.3	3.67	1.95	2.05	4.90	2	3.86	3.52	4.64
	Std. Dev	-	-	-	-		-	0.64	1.13	0.90	0.29	-	0.77	1.05	0.71
NO.10	Mean	25%	23%	34%	13%	5%	69.3	3.00	1.54	4.23	3.83	2	3.42	2.46	3.62
	Std. Dev	-	-	-	-		-	1.00	0.93	0.80	1.06	-	0.84	1.01	0.78
Mean		24%	11%	52%	13%	1%	74.3	4.26	1.41	2.19	4.35	2.4	2.87	2.27	3.17

Wo = woodland, Wa = water bodies, L = lawn, P = pavements, L_{Aeq} = A-weighted sound pressure level, S1 = traffic sounds, S2 = construction sounds, S3 = human sounds, S4 = natural sounds, V = behavioral vitality, E1 = pleasure, E2 = calmness, and E3 = satisfaction.

Figure 4. Measurement point locations and average sound pressure levels.

The overall satisfaction level in Bailuwan Wetland Park is relatively high, with an average of 3.12. The highest satisfaction level is at point 9 (4.14/5), while the lowest is at point 6 (2.21/5). There are four points with satisfaction levels below 3.00: point 2 (2.43/5), point 3 (2.69/5), point 5 (2.82/5), and point 6 (2.21/5). The satisfaction levels for the remaining points range between 3.00 and 4.00.

3.2.1. Descriptive Analysis of Visual Environmental Characteristics

Using a geographic information system (QGis) and space syntax, the proportions of lawn, woodland, water, and pavement elements within the site were calculated, as shown in Figure 5. It was found that lawns are distributed throughout the site, with higher visibility near the viewing platform on the east side of the park. Water bodies are also widely distributed, mainly concentrated in the northern part of the site near point 6, Tinglu Island. Woodland in the western area is scattered, while in the eastern area, it is mainly concentrated in the southern part. A few pavements intersect the site, with higher density and higher pavement grades in the north.

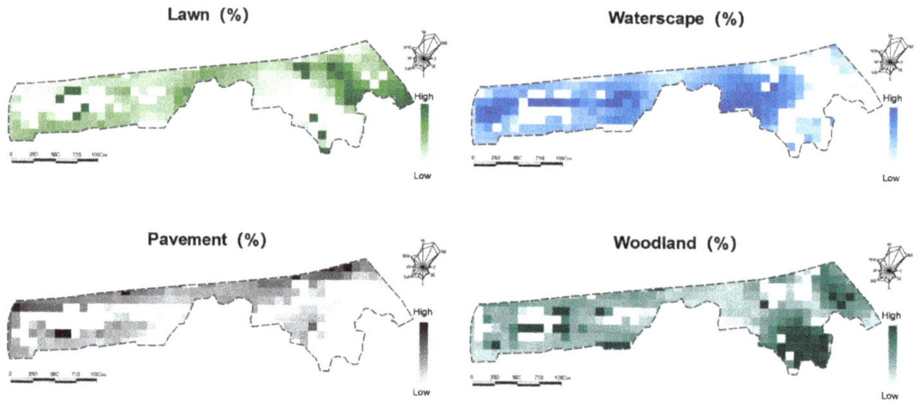

Figure 5. Visibility distribution of visual elements in Bailuwan Wetland Park.

The proportions of visual elements for the 10 measurement points were compared, as shown in Table 2 and Figure 5. The proportion of lawn (mean = 51.54%) was the highest, followed by woodland (mean = 24.02%), pavement (mean = 13.68%), and water (mean = 10.19%) as the least among visual elements. Six measurement points had lawn proportions greater than 50%: point 5 (78%), point 6 (66%), point 2 (65%), point 3 (57%), point 7 (54%), and point 1 (52%). Point 8 (49%) and point 4 (45%) were slightly below 50%, while point 9 and point 10 had the lowest proportions, at 27% and 33%, respectively.

Regarding woodland elements, most points had proportions between 20 and 35%, including point 1 (34%), point 8 (30%), point 4 (28%), point 10 (25%), and point 3 (22%), with only point 9 (57%) exceeding 50%. The points with the least visible woodland elements were point 6 (3%) and point 5 (6%), followed by point 2 (17%) and point 7 (18%).

Water is an important component of the wetland park landscape, and except for point 9, all points had visible water elements. Although water elements were not a design necessity, they played a significant role in psychological perception. The highest visibility for water elements was at point 10, located near Baxian Bridge and adjacent to the water. Points 6, 7, 8, and 4 had water proportions of between 10 and 20%, while the remaining points had less than 10% visibility. There were no large hardscape areas within the site, and the proportion of pavements at each point was similar, fluctuating around 10%. The highest pavement visibility was at points 2, 3, and 9, each with 16%, indicating a dense pavement network. The lowest was point 8, with a pavement grade of only 7%.

3.2.2. Analysis of Auditory Environmental Characteristics

The average sound pressure level across the 10 measurement points in the site was 74.3 dBA. Eight out of these ten points had sound pressure levels exceeding 70 dBA, surpassing the noise standards stipulated by China's "Environmental Quality Standards for Noise" for areas adjacent to expressways. The highest sound pressure level was recorded at point 7 (79.2 dBA), followed by point 8 (78.5 dBA) and point 6 (76.8 dBA). Points 2 (75.5 dBA), 1 (75.5 dBA), 5 (75.5 dBA), and 3 (75.5 dBA) all had sound pressure levels around 75 dBA. Only points 10 and 9, which were farther from the expressway, had decibel values meeting the standards, slightly below 70 dBA, as shown in Table 2 and Figure 6.

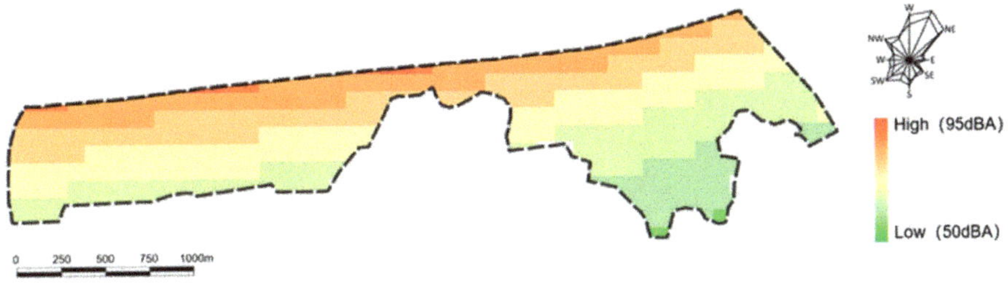

Figure 6. Noise map of Bailuwan Wetland Park.

Based on the traffic flow on the northern side of the site and the measured sound pressure levels at the 10 points, a noise map of the site was generated using the noise prediction model (NPL) [5], as illustrated in Figure 6. The simulation results indicated that the expressway was the primary noise source within the site, with the highest environmental sound pressure levels observed in the northern part, gradually decreasing with distance.

The perception of traffic noise was strong at all 10 measurement points, with perception levels above 3.00. Points 9 (3.67/5) and 10 (3.00/5) exhibited slightly weaker perceptions, whereas point 6 (4.92/5) had the strongest. The perception of mechanical noise was the weakest across all points, with levels below 2.00. As a wetland park, natural sounds were strongly perceived at all points, particularly at points near woodland, such as point 9, which experienced abundant bird calls. The strongest perception of human sounds was at point 10 (4.23/5), located by the water, where traffic noise was minimal and human conversations and laughter were clearly audible (Figure 7).

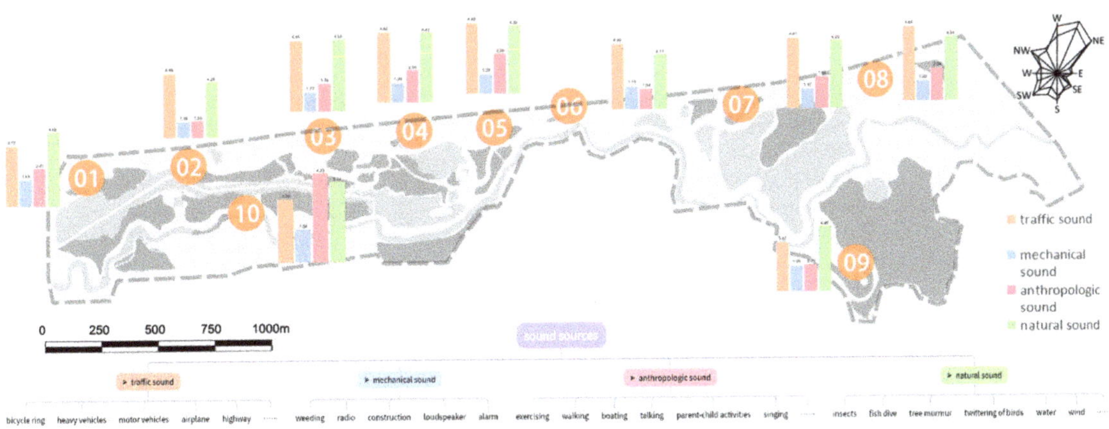

Figure 7. Sound perception map.

3.2.3. Analysis of Behavioral Vitality Characteristics

Based on the distribution of landscape elements within the site, Bailuwan Wetland Park can be divided into four zones, each characterized by its predominant landscape element: the Ecological Conservation Area where forest land is most abundant, the Traffic Activity Area with the highest proportion of roads, the Recreational Area dominated by lawn elements, and the Wetland Protection Area which is richest in water body elements. The area mainly consists of nature reserve areas and wetlands, with an overall lower level of activity, and the average value is 2.4. Open space points 1, 2, 7, 8, 3, and 4 are located in the northwest region of the site, with a behavioral vitality rating of only 3. Points 9 and 10 follow with an activity rating of 2. Point 5 shows minimal human activity, with an activity rating of 0, as depicted in Figure 8.

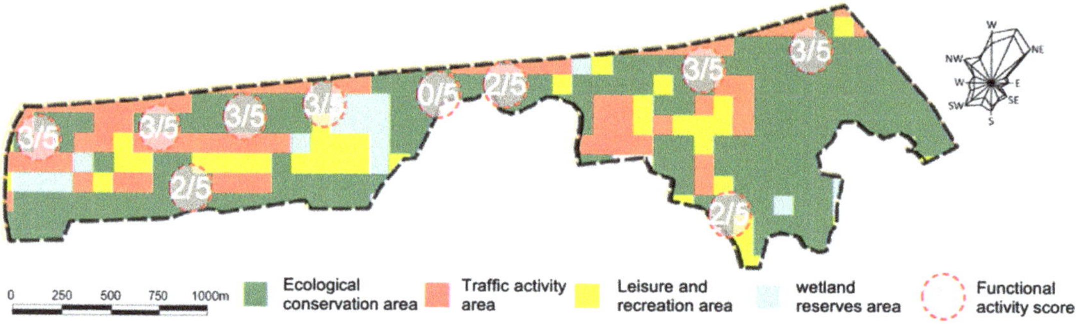

Figure 8. Functional zones and activity intensity map of Bailuwan Wetland Park.

Correlation analysis between site activity intensity and audiovisual environmental characteristics revealed significant associations with woodland ($p < 0.001 < 0.05$), lawn ($p < 0.001 < 0.05$), pavement ($p < 0.001 < 0.05$), and sound pressure level ($p < 0.001 < 0.05$). At the same time, the Shapiro–Wilk test and the Kolmogorov–Smirnov test results for behavioral vitality both have p-values greater than 0.05, indicating that the data conform to a normal distribution. Based on this finding, a multiple regression model was attempted with site activity intensity as the dependent variable and woodland, lawn, and sound pressure levels as independent variables. During model fitting, it was found that woodland elements exhibited significant collinearity with site activity intensity (VIF > 10) and were subsequently removed, resulting in a final predictive model for site activity intensity with an $R^2 = 0.597$ (Equation (1)):

$$V = -16.211 - -0.58L + 0.9P + 0.275L_{Aeq} \quad (1)$$

The model was validated, as shown in Table 3. All independent variables were significant at <0.005, the D-W test result was $1.622 \approx 2$, and VIF values were all less than 10, indicating high model reliability. The model suggests that site activity intensity is negatively correlated with the proportion of lawn in the field of view and positively correlated with pavement visibility and the sound pressure level. In other words, areas with higher greenery exhibit lower activity intensity, whereas areas with better pavement accessibility and livelier sound environments show higher activity intensity.

Table 3. Feasibility assessment of site functional activity fit models.

Model Fit (R^2)	Durbin–Watson Test	Attribute	Estimate B	Standard Error	t-Value	p-Value	VIF
0.597	1.622	lawn	−0.058	0.004	−16.043	<0.001	1.885
		pavement	0.090	0.011	8.115	<0.001	1.898
		lAeq	0.275	0.016	17.127	<0.001	2.785

Mapping the model using QGIS generated the predicted map of site activity intensity (Figure 9). It reveals that areas with high activity intensity are concentrated in the northern side of the site and along the riverbanks, while the utilization of water surfaces is minimal.

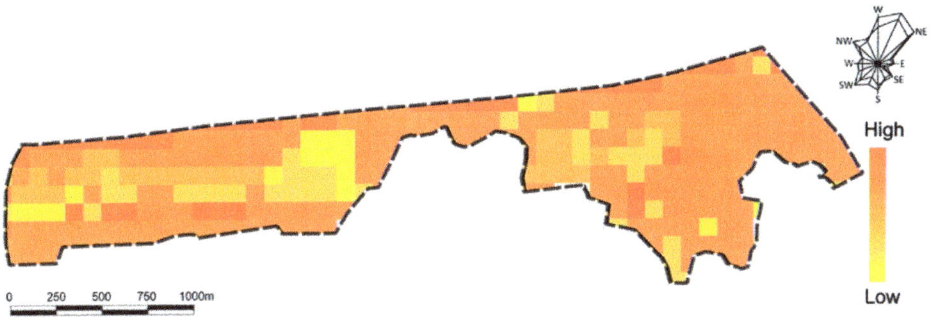

Figure 9. Predicted map of site functionality and activity intensity in Bailuwan Wetland Park.

3.3. Analysis of Factors Influencing Sound Source Perception and Emotional Perception

3.3.1. Analysis of Factors Affecting the Intensity of Sound Source Perception

A correlation analysis was conducted between the perceived intensity of four types of sound sources and visual landscape elements, as well as functional activity levels (see Table 4). The results indicate that the perceived intensity of traffic sounds is significantly positively correlated with the visibility of lawns and pavements and negatively correlated with the visibility of woodland. Pavements with high visibility, often the primary source of traffic sounds, tend to have better accessibility, resulting in stronger perceived traffic noise. Conversely, areas with high lawn visibility are typically more open and lack barriers that could impede noise propagation, leading to a stronger perception of traffic sounds.

Table 4. Correlation analysis of the perceived intensity of sound sources with audiovisual environmental characteristics and behavioral vitality features.

		Wo	Wa	L	P	V
S1	Correlation (Pearson's r)	−0.347 **	−0.093	0.471 **	0.161 *	0.063
	Significant (*p*)	<0.001	0.183	<0.001	0.021	0.366
S2	Correlation (Pearson's r)	0.183 **	−0.023	−0.195 **	0.194 **	−0.062
	Significant (*p*)	0.008	0.747	0.005	0.005	0.376
S3	Correlation (Pearson's r)	0.124	0.340 **	−0.367 **	−0.119	0.235 **
	Significant (*p*)	0.074	<0.001	<0.001	0.088	0.001
S4	Correlation (Pearson's r)	0.204 **	−0.291 **	0.038	−0.240 **	−0.037
	Significant (*p*)	0.003	<0.001	0.588	<0.001	0.591

Wo = woodland, Wa = water bodies, L = lawn, P = pavements, S1 = traffic sounds, S2 = construction sounds, S3 = human sounds, S4 = natural sounds, and V = behavioral vitality. * Significant at 5% level. ** Significant at 1% level.

The perceptibility of construction sounds shows a positive correlation with the visibility ratio of woodland and pavements and a negative correlation with the visibility ratio of lawns. This may be due to construction devices in parks often being located along roads with dense pedestrian traffic or concealed within wooded areas. The perceptibility of human sounds is positively correlated with the visibility of water bodies and the level of behavioral vitality and inversely correlated with the visibility of lawns. In these settings, lawns are generally not favored, and people tend to congregate near water bodies and in areas with higher levels of functional activity.

The perceptibility of natural sounds is primarily related to the visibility of woodland and pavements; increased woodland and reduced pavement visibility are associated with more pronounced natural sounds. Interestingly, the visibility of water bodies, a common source of natural sounds, shows an inverse relationship with the perception of natural sounds in this study. This anomaly may be attributed to the masking effect created by the excessive presence of human sounds around water bodies [50].

3.3.2. Analysis of Factors Influencing Emotional Perception

The analysis indicates that within a certain range of sound pressure levels, human psychological perception of the soundscape is related to the intensity of the sound pressure level. However, when the sound pressure level exceeds this range, the relationship between the two changes. The relationship between the sound pressure level and people's psychological perception of satisfaction at 10 points in the study site is illustrated in Figure 10. Correlation tests revealed that when the sound pressure level is below 77 dB, the correlation coefficient is R = −0.915, with a *p*-value of less than 0.01, indicating a significant negative correlation between the two. Conversely, when the sound pressure level exceeds 77 dB, the correlation coefficient changes to R = 0.985, with a *p*-value of 0.111, which is greater than 0.05, indicating no significant correlation between the psychological perception of satisfaction and decibel values. The perceived satisfaction initially increases and then gradually stabilizes.

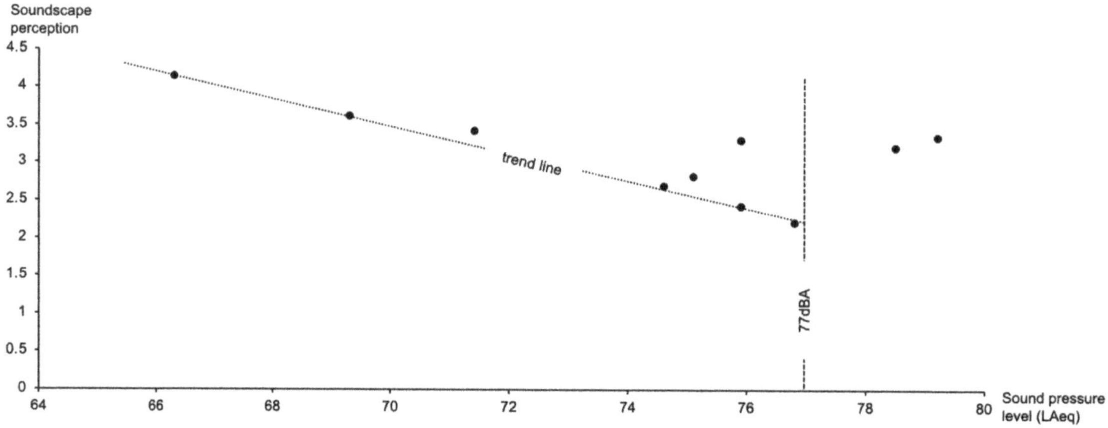

Figure 10. Scatter plot of the relationship between sound pressure level and psychological perception.

Subsequently, a correlation analysis was performed on the other two types of emotions, audiovisual landscape characteristics, and behavioral vitality features. The analysis confirmed that the perception of pleasant emotions (E1) is influenced by visual landscape elements, such as woodlands, lawns, and pavements, and auditory landscape elements, such as traffic noise, human sounds, and natural sounds, as well as the level of behavioral vitality. Specifically, higher visibility of trees and roads, lower visibility of lawns, an increase in human and natural sounds, and more active populations contribute to a higher degree of pleasant emotions.

The perception of calm emotions (E2) is influenced by the visibility of woodlands, lawns, and pavements, the level of behavioral vitality, the perceived intensity of traffic noise, and the A-weighted sound pressure level. A denser woodland, higher road accessibility, fewer lawns, lower sound pressure levels, and reduced functional activity are associated with a greater likelihood of experiencing calm emotions. Details are provided in Table 5.

Table 5. Correlation analysis of emotional perception with audiovisual environmental characteristics and behavioral vitality features.

		Wo	Wa	L	P	S1	S2	S3	S4	L_{Aeq}	V
E1	Correlation (Pearson's r)	0.472 **	0.029	−0.494 **	0.128 *	−0.466 **	0.036	0.276 **	0.130 *	−0.261 **	0.120 *
	Significant (p)	<0.001	0.658	<0.001	0.046	<0.001	0.572	<0.001	0.043	<0.001	0.045
E2	Correlation (Pearson's r)	0.467 **	−0.070	−0.432 **	0.197 **	−0.314 **	0.055	0.055	0.110	−0.075	−0.496 **
	Significant (p)	<0.001	0.275	<0.001	0.002	<0.001	0.390	0.141	0.088	0.244	<0.001

Wo = woodland, Wa = water bodies, L = lawn, P = pavements, S1 = traffic sounds, S2 = construction sounds, S3 = human sounds, S4 = natural sounds, V = behavioral vitality, and L_{Aeq} = A-weighted sound pressure level. * Significant at 5% level. ** Significant at 1% level.

Additionally, a correlation analysis of satisfaction revealed that satisfaction is positively correlated with calmness and pleasure, with a *p*-value of less than 0.001. In summary, the perception of calm emotions is related to decibel levels; when the decibel level is below 77 dB, it can simultaneously affect the psychological perception of satisfaction. However, the variation in pleasant emotions is not related to sound pressure levels. When the decibel level exceeds 77 dB, satisfaction can be enhanced by increasing the sense of pleasure, thereby improving overall perceived satisfaction.

3.4. Emotional Perception Model Prediction

To directly benefit landscape design and perception prediction, we further explored the emotional model equations. After excluding variables with a Variance Inflation Factor (VIF) greater than 10, such as the proportion of lawn elements, and variables with a less significant impact on the dependent variable (e.g., visual area and human sounds), we constructed a model using the perception intensity of pleasant emotions as the dependent variable. The independent variables included lAeq, woodlands, water bodies, traffic noise, and natural sounds, resulting in an R^2 of 0.500 (Equation (2)).

$$E1 = 2.915 -- 0.362S_1 + 0.193S_4 + 0.27Wo + 0.015Wa -- 0.032V \quad (2)$$

All independent variables in this model were statistically significant ($p < 0.05$), with a Durbin–Watson test value of 1.547, approximately 2, and VIF values all less than 10, indicating high model reliability, as shown in Table 6. The model suggests that, in addition to the visibility of woodlands and water bodies and the perceived intensity of natural sounds, higher levels of functional activity and a stronger perception of traffic noise negatively impact the perception of calm emotions.

Table 6. Feasibility test of the emotional perception fitting models.

Emotion	Model Fit (R^2)	Durbin–Watson Test	Attribute	Estimate B	Standard Error	t-Value	*p*-Value	VIF
E1	0.500	1.547	Woodland	0.025	0.402	0.433	<0.001	1.493
			Water bodies	0.015	0.004	0.145	0.014	1.376
			Activity intensity	−0.362	0.006	−0.039	0.049	1.216
			Traffic sound	−0.362	0.047	−0.388	<0.001	1.139
			Natural sound	0.193	0.054	0.182	<0.001	1.074
E2	0.542	1.645	Woodland	0.051	0.005	0.787	<0.001	2.538
			Pavement	−0.040	0.015	−0.188	0.008	2.021
			L_{Aeq}	−0.027	0.022	−0.108	0.023	3.261
			Activity intensity	−0.121	0.065	−0.121	<0.001	1.739

For the calm emotion perception model, the independent variables are woodlands, pavements, behavioral vitality level, and traffic noise, with $R^2 = 0.542$ (Equation (3)). The lawn element was excluded due to a VIF > 10.

$$E2 = 3.771 -- 0.027L_{Aeq} + 0.051Wo -- 0.040P -- 0.121V \quad (3)$$

All independent variables in this model were statistically significant ($p < 0.05$), with a Durbin–Watson test value of 1.645, approximately 2, and VIF values all less than 10, indicating high model reliability, as shown in Table 6. The model indicates that, in addition to the visibility of woodlands, a higher visibility of pavements, increased levels of behavioral vitality, and lAeq negatively affect the perception of calm emotions.

Mapping the two previous models (Figure 11) reveals that the distribution tendencies of the two types of emotions are consistent, exhibiting a trend of being higher in the south and lower in the north. This pattern is in direct contrast to the distribution of negative emotions. People tend to experience more positive emotions around open water surfaces and large areas of woodland.

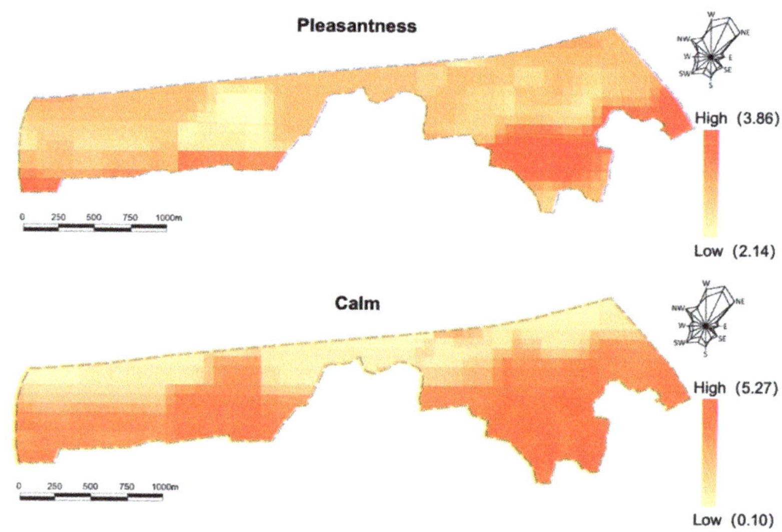

Figure 11. Prediction maps of two emotions in Bailuwan Wetland Park.

4. Discussion

This study, using Bailuwan Wetland Park as a case study, integrates GIS analysis and soundwalk field surveys. From both visual and auditory dimensions, it considers the proportion of visual elements such as lawn, woodland, water bodies, and pavements, as well as the sound pressure level and perceived intensity of four types of sound sources—traffic noise, construction noise, human sounds, and natural sounds—as independent variables. Behavioral vitality levels and emotional perceptions are used as dependent variables to innovatively construct a mid-scale ecological park psychological perception prediction model. This research highlights how visible elements and human activity influence the perceived sound source perception intensity, further contributing to the understanding of soundscape research.

4.1. Discussion on the Influence of Sound Source Perception Intensity

The analysis of factors influencing sound source perception intensity reveals significant correlations between visual landscape elements, functional activity levels, and the perceived intensity of different sound sources. This interplay between visual and auditory elements provides valuable insights for urban park soundscape design, particularly in enhancing acoustic environments to align with user expectations and needs [51,52].

The strong correlation between traffic noise perception and the visibility of grassland and roads highlights the role of visual openness in amplifying undesirable sounds. Grassland areas, being open, offer minimal noise obstruction, leading to a higher perception of traffic noise. In contrast, the negative correlation with tree visibility underscores the effectiveness of vegetative buffers in mitigating noise, aligning with studies on green infrastructure in urban soundscapes [53,54]. Incorporating denser tree cover in high-visibility road areas could reduce the perceived impact of traffic noise.

The relationship between mechanical noise and the visibility of trees and roads suggests that mechanical sources, often near pathways or within wooded areas, are more perceptible. The negative correlation with grassland visibility indicates that open grasslands may be less affected by mechanical noise, possibly due to spatial separation from these sources. This supports research on strategic spatial planning to alter noise perception in urban environments [51,55]. Placing mechanical equipment in less visually prominent locations can reduce their auditory impact.

The correlation between human sound perception and the visibility of water elements and activity levels highlights how people interact with their surroundings. Water bodies, typically seen as tranquil, paradoxically contribute to higher human sound perception due to increased human activity. This masking effect, where natural sounds are overshadowed by human noise, echoes findings on the dual role of water features in soundscape quality [56,57]. Designing buffer zones around water bodies could preserve their natural soundscapes by introducing seating away from the water to minimize noise pollution.

Lastly, the inverse relationship between water visibility and natural sound perception, despite water being a source of natural sounds, underscores the complexity of soundscape design. Prominent water features often attract human activity, diminishing natural sound perception. This can be mitigated by creating quiet zones or enhancing the prominence of natural sounds through landscape design, such as using sound-reflective surfaces or amplifying natural sounds in quieter park areas [43,57].

4.2. Emotional Responses and Audiovisual Landscape Elements

From the three emotional perception models, it was found that the visibility of woodlands and pavements, as well as the perceived intensity of traffic and natural sounds, are significant factors affecting human emotional perception.

Woodlands and natural sounds play a crucial role in fostering positive emotional responses. The presence of trees and greenery has been associated with reduced stress levels and improved mental well-being [9,58]. The natural sounds commonly found in these environments, such as birdsong, rustling leaves, wind, and water, have a significant therapeutic effect on humans [59,60]. These sounds contribute to a sense of tranquility and relaxation, counteracting the stress and anxiety often caused by urban noise. In this study, Bailuwan Wetland Park, located near an expressway, suffers from severe traffic noise pollution. The contrast with the surrounding urban landscape enhances positive feedback when people see woodland landscapes and hear natural sounds. This phenomenon aligns with Attention Restoration Theory, which suggests that natural environments provide restorative experiences that help individuals recover from mental fatigue [9].

The models also reveal that the visibility of water features is significantly correlated with overall satisfaction, although this correlation is weaker in other emotional models. Water is a unique landscape element within the wetland park, present throughout the entire area. During the soundwalk experiments, its presence was consistently perceived. Water features may have a limited immediate visual impact on rapidly changing emotions such as joy, but they subtly influence overall satisfaction through various sensory pathways. The color of the water, its dynamic movements, and the sounds it produces (such as flowing or splashing water) stimulate both visual and auditory perceptions. These sensory experiences contribute to a more profound and enduring effect on people's overall satisfaction with different locations within the park [54,55]. Studies have shown that water features can enhance the aesthetic quality of a landscape, promote relaxation, and improve the overall mood of visitors [56]. The soothing sounds of water can mask unpleasant noises, further enhancing the auditory environment and providing a calming backdrop for park visitors [57].

4.3. Guidance of Landscape Design Based on Sound Pressure Level and Emotional Perception

The analysis identified 77 dB as a pivotal threshold in the emotional response to park environments impacted by traffic noise. Below this threshold, overall satisfaction with the park environment declines as sound pressure levels increase, whereas above 77 dB, no significant relationship was found between satisfaction and sound pressure levels. Therefore, it is recommended that noise mapping should use 77 dB as a key delineation for zoning and targeted environmental improvements.

For park areas where sound pressure levels are below 77 dB, we recommend using a model based on promoting calm emotions to guide design improvements. This model involves controlling sound pressure levels primarily through physical noise reduction measures, such as the installation of sound barriers and protective vegetation. Additionally, perceptual quality can be enhanced by increasing the visibility of woodland areas and reducing the visibility of hardscape elements at the visual level. Furthermore, the model reveals a negative correlation between site activity levels and the perceived calmness of the soundscape. Therefore, the site activity level model can provide further guidance on enhancing the perceived soundscape calmness, such as by appropriately increasing the visibility of grassy areas. These findings offer more refined and feasible strategies for park environment optimization, moving beyond the sole reliance on sound pressure control for improving park quality.

In contrast, for areas where sound pressure levels exceed 77 dB or where an effective reduction in sound pressure is not feasible, a different approach is necessary. In such zones, the focus should shift from sound pressure control to enhancing the perceived pleasantness of the soundscape. This can be achieved by reducing the perception of traffic noise, amplifying the perception of natural sounds, increasing the visual prominence of greenery, and reducing the visibility of paved surfaces. In addition, reducing the activity level can improve people's perceived pleasantness of a soundscape to a certain extent, which reflects the necessity of controlling the intensity of human activities in wetland parks. These strategies indicate that, even in acoustically challenging environments, it is possible to improve the perceptual quality of the soundscape. This dialectical approach highlights the need for landscape design strategies that are responsive to specific acoustic conditions, offering a new perspective on the enhancement of urban park environments.

4.4. Limitations

This study faces several limitations in experimental design and data analysis. Firstly, the experiment was conducted in July, during a period of elevated environmental temperatures, which may have influenced subjective perception evaluations. Future research could address this by conducting studies across different seasons to examine how varying temperatures affect perception. Secondly, the limited number of testing points for measuring soundscape perception within the study site may introduce minor biases into the models. Expanding the number of testing points within and across different parks could improve the accuracy and generalizability of the findings. Thirdly, the study site's landscape is predominantly composed of natural elements, with a lack of hardscape features and built structures. Future research could explore the visual and acoustic effects of diverse landscape compositions, including built environments, to provide a more comprehensive understanding of soundscape perception in different settings.

Moreover, all participants in this study were young adults aged 20–30. This age group may have a higher tolerance for environmental conditions such as temperature and noise, which could potentially influence the generalizability of the findings to other age groups. Future research should consider a broader demographic to assess how age-related differences might impact soundscape perception and environmental tolerance. The inclusion of a more diverse participant pool could enhance the applicability of the results across different population segments.

5. Conclusions

This study, centered on human subjects, employed mathematical modeling to depict the impact mechanisms of environmental audiovisual characteristics on human subjective emotional perception and activities. By integrating this understanding with site-specific environmental information, the study visualized the perceptual quality issues related to audiovisual senses in the studied park. Based on these findings, it proposed leveraging audiovisual interactions to drive sustainable development in parks, offering scientific foundations and flexible strategies to optimize the audiovisual perception quality of urban parks with diverse functionalities. This approach provides a new pathway for the sustainable development of urban green spaces near expressways, aimed at promoting residents' health and well-being. The predictive models developed in this research provide direct design guidance for shaping high-quality urban parks without the need for professional interpretation, thus reducing potential deviations caused by technical and professional language barriers in design objectives. Additionally, they expand the research and practical pathways from single-element perception to the coupling of multisensory elements in landscape perception studies, contributing empirically to the design of urban open spaces that are human-centered and based on multisensory perception. The feasibility of the research methods employed has been validated, suggesting their broad applicability to other cities and types of parks, thereby laying the groundwork for accumulating universally applicable experiences in optimizing urban park audiovisual quality.

Author Contributions: Conceptualization, T.J. and Y.S.; methodology, J.L. and T.J.; validation, T.J., J.L., and Y.S.; formal analysis, J.L.; writing—original draft preparation, T.J. and J.L.; writing—review and editing, T.J. and Y.S.; visualization, J.L.; supervision, Y.S.; project administration, Y.S. All authors have read and agreed to the published version of the manuscript.

Funding: This research was funded by the National Key R&D Program of China [grant number 2023YFC3805303].

Data Availability Statement: The data that support the findings of this study are available from the corresponding author, upon reasonable request.

Conflicts of Interest: The authors declare no conflicts of interest.

References

1. World Health Organization. Healthy Cities Vision. Available online: https://www.who.int/europe/news-room/fact-sheets/item/healthy-cities-vision (accessed on 30 July 2024).
2. U.S. Environmental Protection Agency. Green Infrastructure: Design and Implementation. Available online: https://19january2017snapshot.epa.gov/green-infrastructure/green-infrastructure-design-and-implementation_.html (accessed on 30 July 2024).
3. National Health Service. Healthy New Towns. Available online: https://www.england.nhs.uk/ourwork/innovation/healthy-new-towns/ (accessed on 30 July 2024).
4. National Health Commission. Healthy China 2030 Planning Outline. 2016. Available online: http://www.nhc.gov.cn/xxgk/pages/healthy-china-2030 (accessed on 30 July 2024).
5. Ulrich, R.S. View through a window may influence recovery from surgery. *Science* **1984**, *224*, 420–421. [CrossRef] [PubMed]
6. Kaplan, R.; Kaplan, S. *The Experience of Nature: A Psychological Perspective*; Cambridge University Press: Cambridge, UK, 1989.
7. Maas, J.; Verheij, R.A.; Groenewegen, P.P.; de Vries, S.; Spreeuwenberg, P. Green space, urbanity, and health: How strong is the relation? *J. Epidemiol. Community Health* **2006**, *60*, 587–592. [CrossRef] [PubMed]
8. Barton, J.; Pretty, J. What is the best dose of nature and green exercise for improving mental health? A multi-study analysis. *Environ. Sci. Technol.* **2010**, *44*, 3947–3955. [CrossRef] [PubMed]
9. Kaplan, S. The restorative benefits of nature: Toward an integrative framework. *J. Environ. Psychol.* **1995**, *15*, 169–182. [CrossRef]
10. Hartig, T.; Mang, M.; Evans, G.W. Restorative effects of natural environment experiences. *Environ. Behav.* **1991**, *23*, 3–26. [CrossRef]
11. Hartig, T.; Mitchell, R.; de Vries, S.; Frumkin, H. Nature and health. *Annu. Rev. Public Health* **2014**, *35*, 207–228. [CrossRef]

12. Peters, K.; Elands, B.; Buijs, A. Social interactions in urban parks: Stimulating social cohesion? *Urban For. Urban Green.* **2010**, *9*, 93–100. [CrossRef]
13. Kuo, F.E.; Sullivan, W.C.; Coley, R.L.; Brunson, L. Fertile ground for community: Inner-city neighborhood common spaces. *Am. J. Community Psychol.* **1998**, *26*, 823–851. [CrossRef]
14. Tzoulas, K.; Korpela, K.; Venn, S.; Yli-Pelkonen, V.; Kaźmierczak, A.; Niemela, J.; James, P. Promoting ecosystem and human health in urban areas using Green Infrastructure: A literature review. *Landsc. Urban Plan.* **2007**, *81*, 167–178. [CrossRef]
15. Bowler, D.E.; Buyung-Ali, L.; Knight, T.M.; Pullin, A.S. A systematic review of evidence for the added benefits to health of exposure to natural environments. *BMC Public Health* **2010**, *10*, 456. [CrossRef]
16. WHO. *Urban Green Spaces and Health: A Review of Evidence*; World Health Organization Regional Office for Europe: Copenhagen, Denmark, 2016.
17. EPA. *Green Infrastructure: Managing Wet Weather with Green Infrastructure*; United States Environmental Protection Agency: Washington, DC, USA, 2019.
18. Ulrich, R.S.; Simons, R.F.; Losito, B.D.; Fiorito, E.; Miles, M.A.; Zelson, M. Stress recovery during exposure to natural and urban environments. *J. Environ. Psychol.* **1991**, *11*, 201–230. [CrossRef]
19. White, M.; Smith, A.; Humphryes, K.; Pahl, S.; Snelling, D.; Depledge, M. Blue space: The importance of water for preference, affect, and restorativeness ratings of natural and built scenes. *J. Environ. Psychol.* **2010**, *30*, 482–493. [CrossRef]
20. Berman, M.G.; Jonides, J.; Kaplan, S. The cognitive benefits of interacting with nature. *Psychol. Sci.* **2008**, *19*, 1207–1212. [CrossRef] [PubMed]
21. Alvarsson, J.J.; Wiens, S.; Nilsson, M.E. Stress recovery during exposure to nature sound and environmental noise. *Int. J. Environ. Res. Public Health* **2010**, *7*, 1036–1046. [CrossRef]
22. Krause, B. *The Great Animal Orchestra: Finding the Origins of Music in the World's Wild Places*; Little, Brown and Company: Boston, MA, USA, 2013.
23. Payne, S.R. The production of a perceived restorativeness soundscape scale. *Appl. Acoust.* **2013**, *74*, 255–263. [CrossRef]
24. Zhang, M.; Kang, J. Towards the evaluation, description, and creation of soundscapes in urban open spaces. *Environ. Plan. B Plan. Des.* **2007**, *34*, 68–86. [CrossRef]
25. Jennings, P.; Cain, R. A framework for improving urban soundscapes. *Appl. Acoust.* **2013**, *74*, 293–299. [CrossRef]
26. Gidlöf-Gunnarsson, A.; Öhrström, E. Noise and well-being in urban residential environments: The potential role of perceived availability to nearby green areas. *Landsc. Urban Plan.* **2007**, *83*, 115–126. [CrossRef]
27. Lynch, K. *The Image of the City*; MIT Press: Cambridge, MA, USA, 1960.
28. Appleton, J. *The Experience of Landscape*; Wiley: London, UK, 1975.
29. Brown, A.L.; Muhar, A. An approach to the acoustic design of outdoor space. *J. Environ. Plan. Manag.* **2004**, *47*, 827–842. [CrossRef]
30. Jo, H.I.; Jeon, J.Y. The influence of human-related sounds on the restorative effects of viewing and listening to urban and rural environments. *Sustainability* **2020**, *12*, 2730.
31. Zhang, Y.; Kang, J.; Kang, J. Effects of soundscape on the environmental restoration in urban natural environments. *Noise Control Eng. J.* **2014**, *62*, 480–490.
32. Velarde, M.D.; Fry, G.; Tveit, M.S. Health effects of the physical environment in the urban landscape: A review. *J. Urban Health* **2007**, *84*, 273–284.
33. Brown, A.L.; Kang, J.; Gjestland, T. Towards standardization in soundscape preference assessment. *Appl. Acoust.* **2011**, *72*, 387–392. [CrossRef]
34. Berg, M.; Meekes, J.; Hekmat, A.; van den Berg, A.E. The role of soundscape and visual landscape in health and well-being. *Environ. Behav.* **2014**, *46*, 782–805.
35. Franco, L.S.; Shanahan, D.F.; Fuller, R.A. A review of the benefits of nature experiences: More than meets the eye. *Int. J. Environ. Res. Public Health* **2017**, *14*, 864. [CrossRef]
36. Herzog, T.R.; Black, A.P.; Fountaine, K.A.; Knotts, D.J. Reflection and preference in natural environments. *Environ. Behav.* **2015**, *47*, 113–141.
37. Stokols, D.; Prkachin, K.M.; LaCroix, J.M. Health and well-being in urban environments: An integrative approach. *Environ. Behav.* **2013**, *45*, 471–493.
38. Heng, L.; Hui, X. Investigation of soundscape walks in mountain city parks. In *Proceedings of the 2017 National Acoustics Academic Conference of the Chinese Acoustical Society*; Key Laboratory of Mountainous Town Construction and New Technology, Ministry of Education, College of Architecture and Urban Planning, Chongqing University: Chongqing, China, 2017; pp. 605–606. Available online: https://kns.cnki.net/kcms2/article/abstract?v=-4s28oSk47_vZaY0YFQFXymX-7gWZehTxeCD7FhqfOljK3XxbIUHQI05g4rsd_CLp2om5LE2cd8NoRfEmtlH7MUFdatUH18w5RnJJ_tvafGweLQ2s-CHAE1T-k-wi2ssu6Af31vpQGob5pdNJ7mmvvyvtvnGSf3SOBm08mRK3sW0VNX0jdvhp8s4r0L-CIK4GSTp0d8E0yn3Z0h0I9Z79zfvpbhzmhEqPpgxNTZGFoWWWcYRvn3yRg==&uniplatform=NZKPT&language=CHS (accessed on 5 September 2024).
39. Yin, Y.; Shao, Y.; Lu, H.; Hao, Y.; Jiang, L. Predicting and Visualizing Human Soundscape Perception in Large-Scale Urban Green Spaces: A Case Study of the Chengdu Outer Ring Ecological Zone. *Forests* **2023**, *14*, 1946. [CrossRef]
40. *ISO/TS 12913-2:2018*; Acoustics—Soundscape—Part 2: Data Collection and Reporting Requirements. BSI Standards Publication: London, UK, 2018.

41. Sarwono, J.; Kusdinar, D.; Winarni, S. The effect of music genre on the atmosphere and the buying behavior of café visitors. *J. Cult. Herit. Manag. Sustain. Dev.* **2022**, *12*, 64–76.
42. Russell, J.A. A Circumplex Model of Affect. *J. Personal. Soc. Psychol.* **1980**, *39*, 1161–1178. [CrossRef]
43. Kang, J.; Zhang, M. Semantic differential analysis of the soundscape in urban open public spaces. *Build. Environ.* **2010**, *45*, 150–157. [CrossRef]
44. Mengchi, F. Evaluation and Prediction of Urban Park Soundscape. Master's Thesis, Zhejiang University, Hangzhou, China, 2019.
45. Bahali, S.; Tamer-Bayazit, N. Soundscape Research on the Gezi Park–Tunel Square Route. *Appl. Acoust.* **2017**, *116*, 260–270. [CrossRef]
46. Jo, H.I.; Jeon, J.Y. Overall Environmental Assessment in Urban Parks: Modelling Audio-Visual Interaction with a Structural Equation Model Based on Soundscape and Landscape Indices. *Build. Environ.* **2021**, *204*, 108166. [CrossRef]
47. Cheung, L. Improving visitor management approaches for the changing preferences and behaviours of country park visitors in Hong Kong. *Nat. Resour. Forum* **2013**, *37*, 231–241. [CrossRef]
48. Schultz, T.J. Synthesis of social surveys on noise annoyance. *J. Acoust. Soc. Am.* **1978**, *64*, 377–405. [CrossRef]
49. Fu, X.Y.; Yang, P.; Jiang, S.; Xing, J.X.; Kang, D. Influences on the sense of landscape security by different landscape environment design elements and constitution. *Urban Probl.* **2019**, *9*, 37–44. Available online: https://kns.cnki.net/kcms2/article/abstract?v=-4s28oSk47-TVmzLq-vKbR5BYZ0p4ogzfQCdyd3myQ4zMe3P0rSlESQF7TSTdwgYxsn-3NQa2yc9O0FiN0EtYzQKe0Fb0phLvHXmmeb8yUMyPL-oI6JjL59NAo1v6SHuSnMfXkpvH-NBnPkapQMjuPPhGcv6-s3vlISC3ZBvArn5jspMmbqojwkyz_GGJ4W2rBOS0GwNIISPt31f8Ea11VjEVuiQ7n2JSq2htQ09t-Q=&uniplatform=NZKPT&language=CHS (accessed on 5 September 2024).
50. Koussa, F.; Defrance, J.; Jean, P.; Blanc-Benon, P. Acoustic performance of gabions noise barriers: Numerical and experimental approaches. *Appl. Acoust.* **2013**, *74*, 189–197. [CrossRef]
51. Pheasant, R.; Horoshenkov, K.; Watts, G.; Barrett, B. The acoustic and visual factors influencing the construction of tranquil space in urban and rural environments tranquil spaces-quiet places? *J. Acoust. Soc. Am.* **2010**, *129*, 1210–1217. [CrossRef]
52. Aletta, F.; Kang, J.; Axelsson, Ö. Soundscape descriptors and a conceptual framework for developing predictive soundscape models. *Landsc. Urban Plan.* **2016**, *149*, 65–74. [CrossRef]
53. Van Renterghem, T.; Botteldooren, D. Reducing the acoustical visibility of roads through the application of green roofs and green walls. *Landsc. Urban Plan.* **2009**, *95*, 105–112.
54. Fang, C.F.; Ling, D.L. Investigation of the noise reduction provided by tree belts. *Landsc. Urban Plan.* **2005**, *71*, 29–34. [CrossRef]
55. Nilsson, M.E.; Berglund, B. Soundscape quality in suburban green areas and city parks. *Acta Acust. United Acust.* **2006**, *92*, 903–911.
56. Yang, W.; Kang, J. Acoustic comfort evaluation in urban open public spaces. *Appl. Acoust.* **2005**, *66*, 211–229. [CrossRef]
57. Gozalo, G.R.; Morillas, J.M.B.; González, D.M.; Moraga, P.A.; Vílchez-Gómez, R. Relationship between objective acoustic indices and subjective assessments for the quality of soundscapes. *Appl. Acoust.* **2017**, *116*, 366–374. [CrossRef]
58. Völker, S.; Kistemann, T. The impact of blue space on human health and well-being—Salutogenetic health effects of inland surface waters: A review. *Int. J. Hyg. Environ. Health* **2011**, *214*, 449–460. [CrossRef] [PubMed]
59. Völker, S.; Heiler, A.; Pollmann, T.; Kistemann, T. How does water affect health? Blue space theory and the quality of urban blue spaces. *Water Res.* **2018**, *144*, 236–246. [CrossRef]
60. Annerstedt, M.; Währborg, P. Nature-assisted therapy: Systematic review of controlled and observational studies. *Scand. J. Public Health* **2011**, *39*, 371–388. [CrossRef]

land

MDPI

Article

Towards Healthy and Sustainable Human Settlement: Understanding How Local Communities Perceive and Engage with Spa Tourism Development Initiatives in Rural Areas

Maja Mijatov Ladičorbić [1], Aleksandra S. Dragin [1,2,*], Tamara Surla [1], Aleksandra Tešin [1], Juan Manuel Amezcua-Ogáyar [2], Alberto Calahorro-López [2], Vladimir Stojanović [1], Zrinka Zadel [3], Kristina Košić [1], Olja Munitlak Ivanović [1], Anđelija Ivkov Džigurski [1], Miroslav D. Vujičić [1], Milena Nedeljković Knežević [1], Ljubica Ivanović Bibić [1], Slavica Tomić [4] and Željko Anđelković [5]

1 Faculty of Sciences, Department of Geography, Tourism and Hotel Management, University of Novi Sad, 21000 Novi Sad, Serbia; majam@dgt.uns.ac.rs (M.M.L.); tamara.bozovic@dgt.uns.ac.rs (T.S.); aleksandra.tesin@dgt.uns.ac.rs (A.T.); vladimir.stojanovic@dgt.uns.ac.rs (V.S.); kristina.kosic@dgt.uns.ac.rs (K.K.); olja.mi@dgt.uns.ac.rs (O.M.I.); andjelija.ivkov@dgt.uns.ac.rs (A.I.D.); miroslav.vujicic@dgt.uns.ac.rs (M.D.V.); milena.nedeljkovic@dgt.uns.ac.rs (M.N.K.); ljubica.ivanovic@dgt.uns.ac.rs (L.I.B.)

2 Faculty of Social Sciences and Law, Department of Business Organization, Marketing and Sociology, University of Jaén, Campus Las Lagunillas, 23071 Jaén, Spain; jamezcua@ujaen.es (J.M.A.-O.); acalahor@ujaen.es (A.C.-L.)

3 Faculty of Tourism and Hospitality Management, University of Rijeka, 51410 Opatija, Croatia; zrinka.zadel@fthm.hr

4 Faculty of Economics in Subotica, University of Novi Sad, 24000 Subotica, Serbia; slavica.tomic@ef.uns.ac.rs

5 Faculty of Sciences and Mathematics, Department of Geography, 18000 Niš, Serbia; zeljko.andjelkovic@pmf.edu.rs

* Correspondence: sadragin@gmail.com or aleksandra.dragin@dgt.uns.ac.rs; Tel.: +381-63-8306114

Abstract: This study investigated the intersection of healthy and sustainable human settlement, and land use for spa tourism in rural areas. Recognizing the crucial role of youth in shaping the future of tourism, this research aimed to provide an insight into the young local rural community representatives' perception of sustainable spa tourism development and to identify whether these attitudes are shaped by their socio-demographic characteristics. The applied methodology was the Sustainable Tourism Attitude Scale (SUS-TAS). A total of 254 respondents took part in this research. Using SPSS 30.0.0, the gathered data were subjected to additional analyses based on factor analysis, descriptive statistics, and the general linear model. The results showed a general positive attitude of local community members towards the further development of spa tourism in their communities. The especially important factors were the following: long-term planning of development, environmental protection, negative attitudes towards spa tourism development, economic benefits, as well as community participation. By providing an empirical analysis based on the use of robust tools for measuring youth attitudes, this research offers valuable insights for policymakers, planners, and community leaders into how they can design and implement tourism strategies that align with sustainability principles while fostering local economic development and environmental stewardship. The ultimate goal is to contribute to the broader discussion on developing healthy, sustainable human settlements that balance economic benefits with ecological and social well-being. Synergy/harmony between communities and their natural surroundings is essential for the social, economic, and environmental sustainability of rural human settlements. Mineral springs and spas facilitate sensory experiences for individuals (both tourists and residents) through the utilization of natural resources and the environment.

Keywords: community engagement; health; SUS-TAS scale; sustainable tourism development; young local community perception; rural development

1. Introduction

Health spas, characterized by their use of natural resources and focus on holistic well-being, are increasingly situated in rural areas, where they leverage the unique ecological and cultural attributes of these areas. This trend presents an opportunity to explore the complex interplay between health spas and local community attitudes.

Understanding local community perceptions and engagement is crucial for sustainable management, which contributes towards healthy and sustainable human settlement. As communities increasingly recognize the significance of their ecological and cultural landscapes, the need for informed decision-making becomes apparent. This paper explores the intersection of local perceptions and spa resource management. Understanding local attitudes towards health spa initiatives, particularly among young community members, is vital for fostering sustainable practices and ensuring positive socio-environmental outcomes.

The pursuit of sustainable development has become a central theme in contemporary local and regional planning, particularly in the context of human settlements. Sustainable tourism is increasingly recognized as an essential driver of balanced economic growth, environmental stewardship, and social cohesion [1]. In rural areas, tourism offers a unique opportunity to utilize natural and cultural assets for development while ensuring sustainability [2]. Central to this effort is the concept of land use, which involves managing land resources to support tourism activities on a sustainable basis.

In the changing realm of sustainable tourism, health spas have become important contributors to rural development. These establishments not only offer health and wellness benefits but also intersect significantly with ecological and community dynamics. Health spas in rural areas leverage natural resources and landscapes to deliver sensory and therapeutic experiences, often capitalizing on local ecological features such as mineral springs, scenic vistas, and botanical diversity. The demand for health-focused retreats grows [3–5]. However, their development must be managed carefully to ensure that it aligns with sustainable tourism principles. This paper seeks to outline the perceptions and involvement of the local community as a basis for sustainable management, contributing towards a healthy and sustainable human settlement based on use of natural resources such as thermo-mineral waters.

Spa tourism, a growing segment within the tourism industry, emphasizes health and wellness activities such as therapeutic treatments, relaxation, and physical rejuvenation. This form of tourism not only emphasizes relaxation and wellness but also often intersects with various environmental and socio-economic dimensions of local communities [3–6]. Rural areas, often rich in natural landscapes and resources like mineral springs, provide an ideal setting for spa tourism. This form of tourism can generate economic benefits by attracting visitors seeking wellness experiences, thereby providing a sustainable income source for rural communities [7]. However, the expansion of spa tourism requires careful land use planning to avoid negative environmental impacts and ensure that tourism development adheres to sustainability principles [8].

The attitudes and perceptions of the local population, particularly the youth, play a key role in the success and sustainability of tourism initiatives. Young people are often more environmentally conscious and supportive of sustainable practices [9,10], and mindful of the future [11], making their involvement critical for the long-term viability of tourism projects. Despite this, limited research has specifically examined young residents' attitudes towards land use and spa tourism in rural settings.

To address this gap, it is fundamental to use a robust tool for measuring the attitudes of young local residents towards sustainable spa tourism and land use. The concept of landsenses serves as a foundational element in understanding the emotional and cultural attachments that communities develop towards their environments. This framework integrates various theories and perspectives to illuminate how these attachments influence community perceptions and engagement, particularly in relation to spa resources. In this sense, different theoretical frameworks can be used, such as the following: Place Attachment Theory, Environmental Perception Theory, Cultural Ecology, SUS-TAS, etc. The

Sustainable Tourism Attitude Scale (SUS-TAS) has been used in various contexts to assess local residents' perceptions of tourism impacts [11–14]. However, the applications of this scale often do not account for the specific concerns and perspectives of youth, nor dothey adequately address land use in spa tourism contexts.

The primary goal of this study was to reveal the perceptions of young rural residents about sustainable spa tourism development and to identify which socio-demographic characteristics shaped their attitudes on this issue. In order to assess their opinions, SUS-TAS was applied. In this regard, this study set out to validate the application of SUS-TAS specifically to rural spa tourism, with a focus on the perceptions of young local residents. By doing so, we sought to provide valuable insights that can inform sustainable tourism policies and land use planning, ensuring that spa tourism developments contribute positively to rural economies while preserving environmental integrity and local heritage. The findings from this research bridge the existing gap in the literature and offer practical implications for stakeholders involved in rural development, tourism, and land use planning. More precisely, by providing an empirical analysis based on using robust tools for measuring youth attitudes, this research offers valuable insights for policymakers, spatial planners, and community leaders into how they can design and implement tourism strategies that align with sustainability principles while fostering local ecological–economic development and environmental stewardship.

2. Literature Review

The relationship between local communities and their environments has garnered significant attention in recent years, particularly in the context of landsenses and sustainable development [15] as well as ecosystem services and the benefits for community residents [16,17]. Spa resources, often linked to both ecological and cultural heritage, play a vital role in community identity and well-being. This study drew on the existing literature to examine how perceptions of spa resources influence community engagement and sustainability initiatives.

Health spas often rely on natural landscapes—such as mineral springs, therapeutic muds, and scenic environments—to attract visitors and provide wellness services. While this reliance on natural assets can drive economic benefits and promote health, it also necessitates a careful consideration of ecological impacts and sustainability practices. To ensure that health spas positively contribute to local ecosystems and do not inadvertently degrade the resources they rely on, it is imperative that these resources be managed effectively.

The concept of landsenses—how individuals perceive and value their natural surroundings—provides a valuable framework for understanding local community responses to health spa developments. In rural areas, where the preservation of landscape and natural resources is closely tied to cultural and economic well-being, the attitudes of local community members are particularly significant. This study focused on young community members, a demographic often engaged with both the environmental and economic dimensions of tourism.

2.1. *Towards a Healthy and Sustainable Human Settlement through Spa Tourism*

Tourism can be seen as an approach to sustainable development that covers socio-cultural, economic, and environmental impacts in a rural community [18], and spa tourism in these communities can in such ways affect the community's well-being. The development of spa tourism in villages has become a mechanism for the development of rural areas in many countries. Wellness tourism is highly concentrated in North America, Europe, and Asia Pacific. The top five countries (Unites States, Germany, France, China, Japan) account for 64% of the global market, and the top 20 countries represent 87%. For example, health and wellness tourism in Portugal is considered a strategic product for development that encourages the development of a centuries-old tradition [19]. It has been confirmed that wellness tourists spend more money on travel than the average travelers [20], and the development of spas can lead to a reduction in seasonality and local development in rural areas [21]. With the arrival of visitors, the local population begins to appreciate the environment in which they live more and to become more attached to their place. Thus, in addition to the economic profit, the local population experiences direct consequences such as social and cultural development (interaction with visitors, better quality of life, more space for recreational activities, and improvement of health status) [22]. This creates new prospects for the local population, particularly the youth, encouraging them to remain in rural communities. Consequently, this trend helps curb rural depopulation, enhances the quality of life through improved infrastructure, and preserves the natural and cultural heritage of these areas. Through tourism, the local population not only enhances their own well-being but also contributes to the overall socio-economic progress of underdeveloped regions.

According to Bacsi, Kovács, and Lőke [22], spa tourism is a key growth sector for tourism in Hungary. Development projects in this area help preserve existing jobs, generate new employment opportunities, and create additional value. The resulting multiplier effect can benefit small and medium-sized enterprises in the vicinity of these investments. This, in turn, can play a significant role in reducing spatial and regional disparities.

Based on Draghici et al. [23], studies have shown that health tourism provides an essential part in the ongoing development of local economies and the creation of sustainable management plans for spa tourist destinations. If a successful management and marketing strategy is put in place together with the creation and ongoing enhancement of services, local communities can profit socially and economically from the growth of spa tourism [24,25]. As is the case with Serbia, this is especially important for local economies in developing countries [26,27].

It is important to further research and comprehend the potential benefits of wellness and spa tourism on spa destinations, especially in rural areas [28]. This sector of the tourism market provides good opportunities for the development of rural areas [29], where there is great potential for attracting spa travelers [30,31]. The advantages of health-based travel may be reflected in business models that offer value for the customer, the community, the environment, and the business itself [26]. However, researching residents' views regarding spa tourism development has not been sufficiently explored, although the topic of exploring residents' views on the effects has traditionally been an important topic in tourism studies [32,33].

The integration of health spas into rural areas presents a multifaceted opportunity to advance sustainable tourism while respecting ecological and community dynamics. According to scholars, it is important to explore how perceptions of landscape quality and changes affect local attitudes towards tourism and conservation [34–36].

Understanding the attitudes of young local community members is a prerequisite for assessing the social impact of health spas. Young people often represent a forward-looking perspective on sustainability and development. That is why the topic of young people and their involvement in the local community is increasingly preoccupying scholars [37,38], due to the strong self-awareness of young people about sustainable development and their strong awareness of the future [9–11]. Insights of the young can guide the development of health spas in ways that align with both environmental conservation and community values.

In that context, this research aimed to examine the perspectives of local communities on spa tourism development, using the Sustainable Tourism Attitude Scale (SUS-TAS). The SUS-TAS framework, renowned for its comprehensive approach for assessing the attitudes toward sustainable tourism [14,39,40], provides a robust foundation for exploring the perceptions and involvement of the local communities in spa tourism development initiatives. By applying the SUS-TAS instrument, we sought to identify the multifaceted dimensions of community attitudes, considering the economic, social, cultural, and environmental factors that influence perceptions and preferences.

Based on an examination of local community attitudes, this study adds valuable insights to the academic understanding of sustainable tourism, as well as practical considerations for policymakers, industry stakeholders, and community leaders in fostering a harmonious relationship between spa tourism development and rural local communities.

2.2. Local Community Members' Perception of Sustainable Tourism Development

The intersection of local community dynamics and tourism development has long been a focal point in scholarly research, reflecting the profound impacts that tourism can have on the social, economic, and cultural aspect of communities. As the tourism industry continues to expand globally, understanding the attitudes of the local communities becomes increasingly important for achieving sustainable and mutually beneficial tourism practices [41–43].

Tourism is a multifaceted phenomenon that, when managed effectively, can be a catalyst for positive change, economic growth, and cultural exchange. However, the potential adverse effects, including cultural commodification, environmental degradation, and social disruption, require a careful examination of community attitudes [13]. The SUS-TAS, a well-established tool designed to assess attitudes toward sustainable tourism, provides a comprehensive framework for understanding how local communities perceive and engage with tourism development initiatives [11,12,14].

While studies on community attitudes toward tourism are widely conducted [44–47], there is a noticeable gap in the literature concerning spa tourism and the specific application of the SUS-TAS framework in this context. The existing research often overlooks the nuanced factors that shape community attitudes toward spa tourism development, hindering the formulation of targeted and effective strategies for sustainable coexistence. Through an in-depth analysis, this research aimed to explore the perceptions of the young local community representatives regarding sustainable spa tourism development and to identify which socio-demographic characteristics shaped their attitudes on this issue. This study unravels the intricate interplay between sustainable tourism attitudes and the unique characteristics of spa tourism, contributing valuable knowledge to both academics and practitioners involved in fostering sustainable rural development at the community level and in the context of healthy and sustainable human settlements.

Based on the literature review and identified research gap, the following research questions were established:

RQ1: What is the perception of young residents of rural areas of the sustainable development of spa tourism?

RQ2: Is there a link between the socio-demographic characteristics of young rural residents and their perspectives on the sustainable development of spa tourism?

3. Materials and Methods

3.1. Study Area

This research was conducted in a rural area in the northern part of Serbia (Autonomous Province of Vojvodina) (Figure 1). In that area of 21.506 square kilometers, it was determined that spa tourism can be developed in 50 locations, of which spa complexes can be built in 33 municipalities [28]. The municipalities are predominantly rural in character, and the development of spa tourism would improve the position of those villages, that is, the rural population. "The European Commission identifies rural regions on the basis of urban–rural typology".

Spas along Vojvodina have continuously been adapted to the needs of tourists, so over time, the dominant spa function has been complemented by the tourist function of spas, with more and more wellness content. Thermo-mineral waters can be classified into four categories depending on their temperature: 1. cold mineral waters (equal to the temperature of drinking water), 2. hypothermal waters (20–34 °C), 3. homeothermal waters (34–38 °C), and 4. hyperthermal waters (above 38 °C). The Natural Values consist of several subindicators: therapeutic value of the spa's thermo-mineral waters (related to how many different types of medical conditions and diseases it cures, through different kinds of therapies), utility value of water (for what purposes it can be used—e.g., heating; bottling and drinking; healing—e.g., balneotherapy; recreational purposes—e.g., swimming pools and saunas; electricity production), utilization of thermo-mineral springs (are they used for everything that they could be used for?), number of utilized springs (depending on the number currently used—all of them, half or more, less than half, less than one-quarter) [48]. Although Vojvodina is rich in thermo-mineral sources, they are still not sufficiently utilized in the strategic development of the country's rural areas.

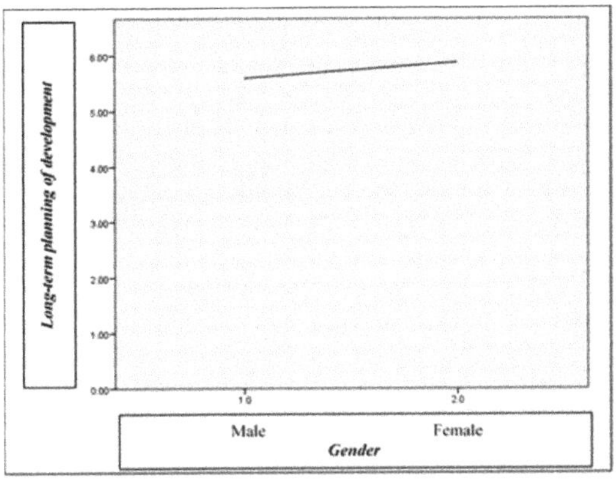

Figure 1. Relationship between long-term planning development and the respondents' gender ($F = 3.942$, $p = 0.049$).

According to foreign and domestic experience, it is estimated that the geothermal waters in Vojvodina, considering its physical–chemical and geothermal characteristics, could be used in the following areas: in agriculture for heating of greenhouses, in cattle and poultry raising for heating farms, in industry as technological warm water, in balneotherapy and sports–recreational-tourist centers, for heating of towns and other buildings, for supplying the population with sanitary warm water, and in fishing and similar. The tourism industry is the most common user of geothermal energy in Vojvodina. This energy is most often used in spas, for healing treatments, recreational purposes, and for the heating of facilities. The use of geothermal energy for heating hotels, housing, or other structures is in the initial stages and very modest considering the availability of resources. The most important and the largest consumers of energy provided by hydrothermal drills are three spas, Spa "Vrdnik", Spa "Junakovic in Apatin, and Spa "Kanjiza" in Kanjiza. These three spas are the most popular among tourists in Vojvodina [49]. The area of spa tourism in Vojvodina is not sufficiently developed in relation to the existing capacities and we cannot be satisfied with the overall results of the development of spa tourism. The large number of sources is not valorized and appears to have unused potential; the tourist offer is one-sided and incomplete; there is a lack of necessary systematic and continued research of a regional and local character, in terms of not only valorization of potential reserves but also in terms of using existing ones; there is an absence of adequate planning and programming documents necessary for development; etc. All of these elements are limiting factors for spa development of tourism in Vojvodina, which should be taken into account in future planning. Spas should be the basis of tourism development in Vojvodina. Today, the degree of utilization of thermo-mineral waters in Vojvodina is unequal, and a transition is required from the primitive, disorganized mode of use to organized use in modern medical centers or in centers for recreation. A more systematic development of the tourist function of spas in Vojvodina will certainly contribute to the preservation and sustainable development of rural areas, as well as the well-being of the local population, which has a chance to benefit from the development of this form of tourism, as well as complementary activities.

3.2. Instrument

This research was conducted using a questionnaire, consisted of questions in two groups. The first group contained several basic socio-demographic questions, including those about a respondent's gender, age, education level, and place of residence. The other group of questions followed the 44 items of the standardized SUS-TAS scale, developed by Choi and Sirakaya [50] and verified by Sirakaya-Turk, Ekinci, and Kaya [39]. In Serbia, this scale, focused on researching the perceptions of the local community members related to sustainable tourism development, was used in terms of the eco-tourism [51,52], and the SUS-TAS scale was mainly implemented in previously conducted studies focusing on protected areas and biodiversity. However, studies conducted in recent years also showed that SUS-TAS scale might be suitable for researching the local community perceptions about sustainable tourism development, based on various tourism contexts, such as the industrial heritage [12,53], while it was also used for researching the stakeholders' perceptions and attitudes about congress tourism development in Serbia [54]. In terms of spa tourism, the short version of the SUS-TAS scale (20 items) was implemented in the context of wellness sustainable tourism development in Thermi (Greece), showing that it might be considered a well-established and reliable quantitative tool [55]. Such findings are supported by the fact that spa tourism development is primarily focused on natural resources (the context in which the SUS-TAS scale is mainly used). Yet, there is a gap in the literature regarding use of the SUS-TAS scale to research the attitudes of the local community members about sustainable development within spa destinations in Serbia, especially among the younger generations, who are considered the carriers of future tourism development. Accordingly, this study was focused on implementing the entire SUS-TAS scale (44 items), not the shorter version (as in the case of Thermi, Greece), which raised the importance of this research. By utilizing the SUS-TAS scale in this study, we aimed to unravel the intricate layers of community attitudes, considering environmental stewardship, social involvement, and economic considerations.

3.3. Sample

Our 254 respondents were 69% females and 31% males. The sample contained 59.4% of respondents aged up to 20 years and 40.6% aged between 21 and 30 years. According to the obtained level of education, the majority of them gained a high school education degree, at 94.4%, while a minority of them finished grammar school (1.2%) or gained a higher university degree (4.4%). In terms of the place of residence, the sample contained respondents from all across Province of Vojvodina, from 64 rural settlements.

In this study, the focus was on representatives of Generation Z, those born between 1995 and 2015 [10], who represent future business leaders. In this view, it can be said that representatives of Generation Z will take over the leading positions, influence business [9,10,56–58], and direct the rural development of their local communities. Another reason was that without young people, especially women, there is no survival of the village [9]. Their attitude towards the economic development of rural areas is therefore highly important to take into account. The classification of regions was determined by identifying the population in rural grid cells (all cells outside of urban clusters) and their proportions. Predominantly rural regions were defined as those in which more than half of the population lived in rural grid cells [59].

3.4. Data Collection Procedure

Data were collected by using a standard pen and paper procedure in the period between October 2022 and January 2023. Respondents answered the questions, or more precisely they expressed their attitudes on the SUS-TAS items, by choosing one answer on the scale between 1 (not important at all) and 7 (completely important), except in the case of the questions regarding the respondents' socio-demographic profile. Respondents were informed that involvement in this study was anonymous and on a voluntary basis, and that their answers would be analyzed for scientific purposes only. Collected data were further analyzed by using SPSS, on the basis of descriptive statistics, factor analysis, and the general linear model.

4. Results

The main component analysis was conducted for the entire SUS-TAS scale, including the 44 items about the local community members' attitudes towards sustainable spa tourism development. The recorded value of the Kaiser–Meyer–Olkin indicator was 0.901, which exceeded the recommended level of 0.6. Bartlett's test of sphericity reached statistical significance ($p = 0.000$) and justified the application of the principal component analysis for the purpose of this research. The principal component analysis revealed the presence of five components with values over 1, explaining 16%, 12%, 11.1%, 10%, and 8.7% of the variance. After the extraction of factors, Varimax rotation with Kaiser Normalization was implemented. One item was excluded from the analyses, due to its low item loading ('It is sometimes acceptable to exclude the local community representatives from the process of making decisions on tourism development'), which resulted in a model with 43 items grouped into five factors explained 57.8% of the total variance. The identified factors are presented in Table 1: *long-term planning of development (1), environmental protection (2), negative attitude towards spa tourism development (3), economic benefits (4) and community participation (5).* A detailed explanation of the factors is provided below.

Table 1. Local community members' perceptions of sustainable spa tourism development (SUS-TAS)—factor analysis.

Rotated Component Matrix [a]					
	Component				
ά = 0.915	Long-Term Planning of Development ά = 0.925	Environmental Protection ά = 0.882	Negative Attitude Towards Spa Tourism Development ά = 0.893	Economic Benefits ά = 0.907	Community Participation ά = 0.816
I believe that a good planning coordination is needed for spa tourism development.	0.757				
Spa tourism development plans should be constantly updated.	0.739				
I believe that a long-term direction should be taken when planning the development of spa tourism.	0.735				
Visitors' satisfaction must be monitored.	0.722				
I believe that successful spa planning management requires an advanced planning strategy.	0.722				
The tourism sector must provide a quality tourism experience for the future visitors.	0.704				
Spa tourism companies have a responsibility to meet the needs of the visitors.	0.682				
The spa tourism industry must have a plan for the future.	0.666				
The spa tourism sector must contribute to development of the local community members' funds.	0.646				
Members of the local community should be given more opportunities to invest in spa tourism development in rural areas.	0.544				
Spa tourism companies should strive to have at least half of the total number of employees from the local community.	0.530				
The tourism sector should use at least half of the goods and services from the local population.	0.492				
The attractiveness of the local community is a basic element of interest to spa visitors.	0.387				

Table 1. *Cont.*

Rotated Component Matrix [a]					
	Component				
ά = 0.915	Long-Term Planning of Development ά = 0.925	Environmental Protection ά = 0.882	Negative Attitude Towards Spa Tourism Development ά = 0.893	Economic Benefits ά = 0.907	Community Participation ά = 0.816
Tourism must protect the environment of the local community.		0.776			
Spa tourism should be developed in accordance with the natural and cultural environment.		0.773			
Spa tourism development must promote a positive attitude towards the environment among all participants in tourism.		0.751			
Proper development of spa tourism means that plant and animal species and their habitats are protected every moment.		0.697			
Diversity of nature must be valued and protected.		0.695			
The environment should be protected now, as well as in the future.		0.666			
I believe that spa tourism must improve the environment for future generations.		0.649			
Laws are required in order to reduce the negative impacts of spa tourism development.		0.600			
I think that development of spa tourism would lead to an intensification of efforts aimed at preserving the environment.		0.408			
I believe that my quality of my life could be threatened with the development of spa tourism.			0.806		
I do not feel comfortable or welcome in the local tourist facilities due to the spa visitors.			0.802		
I think that spa tourists could overuse the tourist resources.			0.790		
I believe that interpersonal relations in my community could be ruined due to the development of spa tourism.			0.772		
I am often irritated by the representation of spa tourism in my community.			0.759		
I think that arrival of spa tourists in my local community could jeopardize my quality of life.			0.723		
I think that development of spa tourism could lead to crowds in my local environment.			0.685		
I think that spa tourism will develop too fast in my local environment.			0.638		
I believe that spa tourism could provide a strong economic contribution to the local community in rural areas.				0.793	
I believe that development of spa tourism is good for the local economy.				0.776	
I believe that spa tourism could provide new income to our local community.				0.764	
Spa tourism could enable development of the local economic diversity.				0.749	
I believe that spa tourism could bring significant tax revenues to the local self-government.				0.733	
Spa tourism could provide access to new markets for our local products.				0.701	
Spa tourism could also provide benefits for the other activities in the local community.				0.475	

Table 1. *Cont.*

Rotated Component Matrix [a]					
	Component				
ά = 0.915	**Long-Term Planning of Development ά = 0.925**	**Environmental Protection ά = 0.882**	**Negative Attitude Towards Spa Tourism Development ά = 0.893**	**Economic Benefits ά = 0.907**	**Community Participation ά = 0.816**
I believe that decisions about spa tourism must be made by all representatives of the local community, regardless of what they do.					0.736
I believe that the involvement of the local community representatives in a decision-making process regarding spa tourism development is a necessary aspect of successful development of this activity.					0.637
I think that members of the local community need to be encouraged to take the lead in spa tourism planning committees.					0.612
Representatives of the local community should have the opportunity to be involved in the process of spa tourism development and management.					0.589
The sector of spa tourism must respect the values recognized by all members of the local community.					0.582
Members of the local community should have a significant share of benefits (incomes) from spa tourism.					0.493

Extraction Method: Principal Component Analysis. Rotation Method: Varimax with Kaiser Normalization.
[a] Rotation converged in 7 iterations.

The first factor, termed long-term planning of development, highlighted the importance of good planning coordination for further spa tourism development, which should be constantly updated and in line with an advanced planning strategy of rural areas. Young local community members also believe that plans for future development of spa tourism require monitoring visitors' satisfaction, as the tourism sector must provide a quality tourism experience for the future visitors, together with all spa tourism companies involved in such development. Zheng et al. (2022) suggest ways to improve visitors' perception of the visited area from the perspective of the ecology of landsenses. The most important indicators are related to vision and touch, including 'vision of plants', 'vision of water', 'feel of sunlight', 'touch of roads', and 'sound of animals' [60], which are characteristic of spa complexes.

Respondents in this study also believe that the attractiveness of the local community in rural areas is a basic element of interest to spa visitors, which is reflected in their attitudes that members of the local community should be given more opportunities to invest in spa tourism development. The tourism sector should strive to have at least half of the total number of employees from the local community and it should use at least half of the goods and services from the local population. Altogether, these could contribute to development of the local community members' funds in rural areas, achieved via the sustainable development of spa tourism.

The second factor is labeled environmental protection. The items grouped within this factor indicate the respondents' perception that further development of spa tourism in rural areas must protect the environment of the local community. Respondents also believe that spa tourism should be developed in accordance with the natural and cultural environment, and that it should promote a positive attitude towards the environment among all participants in tourism. Proper development of spa tourism in rural areas means that plant and animal species and their habitats are protected every moment, or more precisely that the diversity of nature must be valued and protected now, as well as in the future, in order to improve the environment for future generations. This is supported by the findings of the study entitled "Visitors' perception based on the five physical senses of the ecosystem services of urban parks from the perspective of the ecology of the land senses", which states how important the senses of the land are, which relate to plants, water, sunlight, and animals [60].

The following factor, negative attitude towards spa tourism development, indicated the respondents' belief that the quality of their lives in rural areas could be threatened by the development of spa tourism, based on the crowds of tourists. In addition, the local community members often do not feel comfortable or welcome in the local tourist facilities due to the spa visitors. They also believe that interpersonal relations in their community could be ruined due to development of spa tourism, especially in the case of development too fast.

The principal component analysis identified another factor, labeled economic benefits. It seems that respondents are aware of the fact that spa tourism could provide a strong economic contribution to the local community in rural areas, by enabling opportunities for new incomes, as well as significant tax revenues to the local self-government. Respondents also believe that spa tourism could provide access to new markets for their local products, which could enable the development of local economic diversity and provide benefits for other activities. In addition, this segment is also important in terms of the perception of the quality of life of young people in the local environment. If their current place of residence allows them to do what they want and achieve their goals, they are more likely to form a functional relationship, stay, and not move [61].

The last factor is termed community participation, which puts the focus on the fact that representatives of the local community from rural areas should have the opportunity to be involved in the process of spa tourism development and management and have a significant share of the benefits (income) from spa tourism. They also believe that decisions about spa tourism must be made by all representatives of the local community, regardless of what they do. They should be encouraged to take the lead in spa tourism planning committees, in order to be involved in a decision-making process regarding spa tourism development in rural areas. Finally, the local community members believe that the sector of spa tourism must respect the values recognized by all members of the local community. This aligns with the study entitled "A Method for Assessing Residents' Perceptions of Their Community Based on Landsenses Ecology", where it is stated that competition for resources due to tourism development can negatively affect residents' psychological perceptions of their community. However, when residents are actively involved in decision-making and benefit distribution in tourism planning, their perceptions of tourism's impact can be enhanced [62].

Besides the structure of these factors, it is also important to consider the scale reliability, which is presented in Table 1. As can be seen, Cronbach's alpha coefficient for the entire scale (all 44 items) was 0.910, which indicated a satisfactory reliability when implementing the SUS-TAS scale within this study. After excluding one item from further analyses, the value increased to 0.915 (for 43 items). On an individual basis, the highest value of Cronbach's alpha coefficient was recorded for the factor labeled long-term planning of development (0.925), while the lowest but still satisfactory value of Cronbach's alpha coefficient was recorded for the factor community participation (0.816).

The following analyses were focused on calculating the mean values of each factor. The research results, presented in Table 2, showed that the highest mean value was recorded for the factor environmental protection (M = 6.25), while slightly lower mean values were recorded for the factors long-term planning of development (M = 5.79), economic benefits (M = 5.75), and community participation (M = 5.11). We found an encouraging result in the case of the lowest mean value, which was recorded for negative attitudes towards spa tourism development (M = 2.81).

Further analyses were conducted on the basis of the multivariate general linear model, in order to research whether there were significant differences in the identified factors according to the respondents' socio-demographic characteristics. The research results did not point to significant relations between the identified factors and age. On the other hand, significant results were identified in the case of relations between the respondents' perception of *long-term planning development* and their age (F = 3.942, p = 0.049), as well as in the case of *negative attitudes towards spa tourism development* and the respondents' education level (F = 5.466, p = 0.005), as can be seen in Table 3.

Table 2. Attitudes of the local community members towards sustainable spa tourism development (mean values).

Factor	Minimum	Maximum	Mean Value	Standard Deviation
Long-term planning of development	1	7	5.79	0.95
Environmental protection	1	7	6.25	0.81
Negative attitudes towards spa tourism development	1	7	2.81	1.29
Economic benefits	1	7	5.75	1.07
Community participation	1	7	5.11	1,11

Table 3. Perception of the SUS-TAS factors according to the respondents' socio-demographic characteristics—general linear model results.

Source		F	Sig.
Gender	Long-term planning of development	3.942	0.049
	Environmental protection	0.159	0.691
	Negative attitudes towards spa tourism development	0.476	0.491
	Economic benefit	3.129	0.078
	Community participation	0.027	0.870
Age	Long-term planning of development	1.727	0.105
	Environmental protection	1.408	0.204
	Negative attitudes towards spa tourism development	0.730	0.646
	Economic benefit	10.444	0.190
	Community participation	0.343	0.933
Education	Long-term planning of development	1.324	0.268
	Environmental protection	0.065	0.937
	Negative attitudes towards spa tourism development	5.466	0.005
	Economic benefit	1.633	0.198
	Community participation	1.776	0.172

Long-term planning of development. R Squared = 0.081 (Adjusted R Squared = 0.033). Environmental protection. R Squared = 0.050 (Adjusted R Squared = 0.001). Negative attitudes towards spa tourism development. R Squared = 0.079 (Adjusted R Squared = 0.031). Economic benefit. R Squared = 0.079 (Adjusted R Squared = 0.032).

More precisely, there was a significant relation between the respondents' perception of the *long-term planning of development* and their age (F = 3.942, p = 0.049) (see Figure 1). Although this difference was small, it was statistically significant and pointed to the fact that female respondents' perception of the aforementioned factor was slightly higher than that of male respondents.

Finally, Figure 2 points to significant research results in the case of the respondents' perception of *negative attitudes towards spa tourism development* on the basis of their education level. Perception of this factor was highest among those with a high school or university education level.

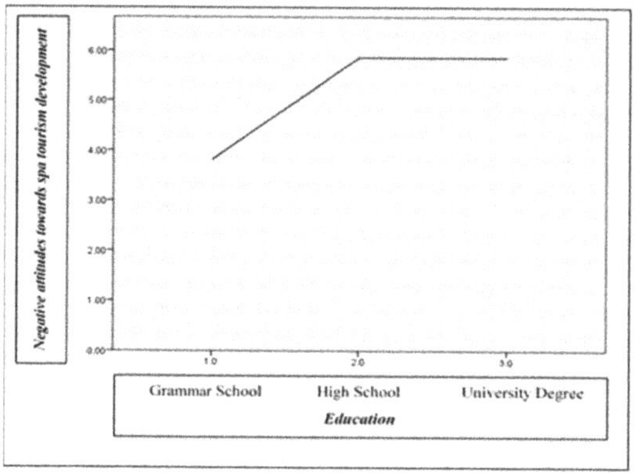

Figure 2. Relationship between negative attitudes towards spa tourism development and the respondents' education level (F = 5.466, p = 0.005).

5. Discussion

The integration of health spas into rural areas presents a multifaceted opportunity to advance sustainable tourism while respecting ecological and community dynamics.

Spa tourism development initiatives, which typically involve the construction and promotion of wellness facilities and related services, hold the potential to contribute to both economic growth and the enhancement of local quality of life. However, the success and sustainability of these initiatives are contingent upon the perceptions and engagement of the communities in which they are implemented. Local residents' attitudes towards spa tourism can significantly influence the effectiveness of these initiatives and their long-term benefits.

This paper aims to bridge a critical gap in the current literature by outlining how local communities perceive and engage with spa tourism development. For this purpose, we used questionnaires, which are a valuable tool supported by established technical standards and previous research [15].

The main results of this study on the attitudes of the local community members towards sustainable spa tourism development in rural areas, collected using the SUS-TAS scale, were deduced through the principal component analysis based on 44 items. The results identified five key factors: long-term planning of development, environmental protection, negative attitude towards spa tourism development, economic benefits, and community participation, indicating the main aspects of local community members' perceptions. We propose that a focus on these is required for further sustainable development of spa tourism.

The first factor, long-term planning of development, emphasizes the importance of adequate planning coordination, constant updates, and an advanced planning strategy in rural areas. The second, environmental protection, reflects the respondents' attitude that spa tourism development in rural areas should be in line with natural and cultural environments, promoting a positive attitude towards the environment. The third factor, termed negative attitudes towards spa tourism development, indicates the local community members' concerns regarding the potential threats to the quality of their lives in rural areas, discomfort in local facilities due to spa visitors, and potential overuse of tourist resources. The fourth factor, economic benefits, recognizes the potential economic contribution of spa tourism in rural areas, including new income, tax revenues, and access to new markets for the local products. The last factor, community participation, highlights the importance of involving the local community members in the decision-making process for rural areas, thus ensuring them a significant share of the benefits and respecting the values that are accepted within the community.

One of the main actors in the economy of tourism is the local community [63], and it is encouraging to see that the respondents generally have a positive attitude toward the growth of spa tourism.

In the research results, the perception of the environmental protection factor (M = 6.25) had the highest mean value. The literature emphasizes the significance of educating the local population on this topic [64,65], and these data demonstrate the high awareness of sustainability and the desire to protect the environment among the local population, which is considered vital.

The surveyed population largely believes that long-term development planning is needed (M = 5.79) in order for spa tourism to have positive effects on the rural destination. According to research conducted by the authors Kim et al. [66], which highlights the significance of active participation and involvement of local communities for the successful development of tourism, spa tourism destinations should involve the local population more in their activities (employment, product sales, etc.).

Our findings reveal that local community perceptions significantly impact engagement with spa resources. Communities with positive perceptions tend to participate more actively in sustainability efforts, leading to better management outcomes. Conversely, negative perceptions can result in resistance to initiatives perceived as harmful or exploitative. The discussion emphasizes the importance of integrating local knowledge and cultural values into spa resource management, suggesting that a more inclusive approach can enhance community engagement and lead to healthier, more sustainable settlements.

According to research of Petrović et al. [67], the local community's level of participation and response to the growth of tourism frequently depend on the individual benefits that tourism provides for the community. Similarly, in this case, despite their initial disapproval, the residents who depend on tourism for their income are now in favor of the growth of spa tourism in their community [32]. It follows that the high mean value of the economic benefit factor (M = 5.75) is justified.

Since the local population is an integral part of the tourism product, negative attitudes toward tourism development can keep visitors, searching for opportunities for learning and participation, from having authentic experiences that meet their expectations [68]. In this regard, it is encouraging that the factor "negative attitude towards the development of spa tourism" (M = 2.81) had the lowest mean value.

Similarly to previous research that indicated respondents' socio-demographic characteristics have a major impact on how they perceive tourism in their areas [69], certain relationships were found in the present research. We found significant relationships between the local community members' age bracket and their perception of long-term planning of development, as well as between the level of the respondents' education and their negative attitude towards spa tourism development. More precisely, female respondents showed a slightly higher perception of long-term planning of development, while respondents with high school or university education had a higher perception of negative attitudes towards spa tourism development.

6. Conclusions

The synergy of humans and nature is a crucial principle for sustainable management of the environment and natural resources. Successful management of spas and mineral springs contributes to the healthy and sustainable development of rural settlements, enabling multiple benefits for the community.

This study underscores the critical role of local community perceptions in the sustainable management of spa resources. By fostering engagement and incorporating ecological and cultural contexts, stakeholders can create strategies that not only protect these resources but also promote community well-being. Our findings contribute to the broader discourse on sustainable human settlements, highlighting the necessity of considering local voices in environmental decision-making processes.

Based on the examination of the local communities' attitudes, this research contributes valuable insights to the academic understanding of sustainable tourism and the practical considerations that guide policymakers, industry stakeholders, and community leaders in fostering a harmonious relationship between spa tourism development and local communities. The main aim of this paper was to provide an insight into the perceptions of the young local community representatives from rural areas regarding sustainable spa tourism development and to identify which socio-demographic characteristics shaped their attitudes on this issue. This research provides a valuable insight into the complex perceptions of the local community members from rural areas of sustainable spa tourism development, emphasizing the key factors and further implications for planning and implementing initiatives, environmental protection, gaining economic benefits, and community involvement. This paper offers both theoretical and practical contributions to the field. Theoretically, it enriches the understanding of how local communities perceive and engage with spa tourism development initiatives in rural areas, contributing towards healthy and sustainable human settlement. More precisely, this research contributes to a better understanding of the local community members' attitudes towards sustainable development of spa tourism in rural areas. While the SUS-TAS scale was widely implemented in studies considering nature preservation and the development of tourism based on nature, this study enables an insight into the local communities' perceptions and their involvement in another context, considering spa tourism development. Practically, it provides actionable insights for policymakers and practitioners, emphasizing the need for inclusive strategies that recognize the unique perceptions of local communities. In addition, this study sampled young local community representatives, those belonging to Generation Z, who will soon approach the age of involvement in the tourism business environment and thus will be in a position to manage its further development. In regard to the main practical implications, the research results might find the scope for practical implementation in policymakers', industry stakeholders' and community leaders' fields of action, supporting a harmonious and balanced relationship between spa tourism development and the local communities' perceptions and behavior. By bridging theory and practice, we hope to inspire more effective and sustainable approaches to managing spa resources, contributing towards healthy human settlements. More precisely, the research results could serve as a data basis for creating a guideline that will increase the local community members' acceptance and their involvement in spa tourism development in rural areas on a sustainable basis.

Besides the main theoretical contribution and practical implication, this research also has several limitations, which lead us to make recommendations for future research, which could focus on different age groups or a dominant education level within the sample. Such additional research might gain results pointing to the fact that there are significant relationships between the respondents' gender and education level, on the one hand, and their perception of sustainable development of spa tourism in rural areas, on the other. Moreover, it could be useful to replicate this study in similar rural areas in other countries. Plus, while this research was focused on identifying significant differences in perceptions of sustainable tourism development in spa areas based on the respondents' socio-demographic characteristics, further studies could consider the other aspects that might cause potential differences, such as the respondents' occupations, spatial distance from the tourism core area, tourism development characteristics, tourism development benefit expectation, etc. Future research could also be focused on adding the constructs of place attachment and well-being, which is greatly relevant as Serbia is facing the problem of depopulation, i.e., the emigration of young people, especially from rural areas. Our intention will be to research place bonding, place identity, as well as place dependence together with perceptions of sustainable tourism development, in order to determine the key factors in young people staying in or leaving the rural area. Future studies could increase the proportion of respondents under 30 years old in the surveyed rural settlements, thus emphasizing the importance of research on young people's attitudes towards tourism development in more detail. The main findings of this study might also instruct the development of future study programs for higher education in tourism, demography, agriculture, and the economy that are focused on sustainable development and instigating social, environmental, economic, and cultural regeneration.

Author Contributions: Conceptualization, M.M.L., A.S.D., and T.S.; data curation, M.M.L.; formal analysis, M.M.L. and A.S.D.; funding acquisition, A.S.D., A.C.-L., Z.Z., Ž.A., and J.M.A.-O.; investigation, M.M.L., O.M.I., M.N.K., and V.S.; methodology, M.M.L. and A.S.D.; resources, A.S.D., Z.Z., and J.M.A.-O.; supervision, M.M.L.; A.S.D., and Z.Z.; validation, M.M.L., A.S.D., and T.S.; visualization, M.M.L., A.S.D., T.S., and A.C.-L.; writing—original draft, M.M.L., A.S.D., T.S., A.C.-L., and Z.Z.; writing—review and editing, M.M.L., A.S.D., T.S., A.T., O.M.I., K.K., A.I.D., M.D.V., L.I.B., and S.T.; project administration, K.K. All authors have read and agreed to the published version of the manuscript.

Funding: This research was funded by the Autonomous Province of Vojvodina, Provincial Secretariat for Higher Education and Scientific-Research Activity, Program 0201, with the project title "Research of the entrepreneurial potentials among the local population for using the thermo-mineral water resources of Vojvodina", registration number: 142-451-3467/2023-02 (2021–2024). The authors are also grateful for the financial support of the Ministry of Science, Technological Development and Innovation of the Republic of Serbia (Grant Nos. 451-03-66/2024-03/200125 and 451-03-65/2024-03/200125).

Data Availability Statement: Dataset available on request from the authors.

Conflicts of Interest: The authors declare no conflicts of interest. The funders had no role in the design of this study; in the collection, analyses, or interpretation of data; in the writing of the manuscript; or in the decision to publish the results.

References

1. Edgell Sr, D.L. *Managing Sustainable Tourism: A Legacy for the Future*; Routledge: London, UK, 2019. [CrossRef]
2. Ursache, M. Tourism–significant driver shaping a destinations heritage. *Procedia-Soc. Behav. Sci.* **2015**, *188*, 130–137. [CrossRef]
3. Szromek, A.R.; Wybrańczyk, K. Proposal of value for customer of spas: Expectations of spa patients and tourist in polish spas. *Sustainability* **2019**, *11*, 3598. [CrossRef]
4. Baloglu, S.; Busser, J.; Cain, L. Impact of experience on emotional well-being and loyalty. *J. Hosp. Mark. Manag.* **2019**, *28*, 427–445. [CrossRef]
5. Koskinen, V. Spa tourism as a part of ageing well. *Int. J. Spa Wellness* **2019**, *2*, 18–34. [CrossRef]
6. Singh, R.; Manhas, P.S.; Mir, M.A.; Quintela, J.A. Investigating the relationship between experience, well-being, and loyalty: A study of wellness tourists. *Adm. Sci.* **2022**, *12*, 98. [CrossRef]
7. Goulding, R.; Horan, E.; Tozzi, L. The importance of sustainable tourism in reversing the trend in the economic downturn and population decline of rural communities. *Rev. Tur. Patrim. Cult.* **2014**, *12*, 549–563. Available online: https://www.pasosonline.org/Publicados/12314/PASOS38.pdf#page=47 (accessed on 1 September 2024). [CrossRef]
8. Lobosco, G. Sizing Landscapes: A Scenario-Based Approach Addressing Landscape Changes due to Infrastructure Developments, Learning from Touristic Contexts. Ph.D. Thesis, Polis University, Tiranë, Albania, 2017. Available online: https://hdl.handle.net/11392/2487916 (accessed on 1 September 2024).
9. Gaidhani, S.; Arora, L.; Sharma, B.K. Understanding the attitude of generation Z towards workplace. *Int. J. Manag. Technol. Eng.* **2019**, *9*, 2804–2812.
10. Dwidienawati, D.; Abdinagoro, S.B.; Gandasari, D.; Tjahjana, D. Do generation Y and Z really concern about environmental issues? *IOP Conf. Ser. Earth Environ. Sci.* **2021**, *729*, 012137. Available online: https://iopscience.iop.org/article/10.1088/1755-1315/729/1/012137/pdf (accessed on 1 September 2024). [CrossRef]
11. Dragin, A.S.; Majstorović, N.; Janičić, B.; Mijatov, M.B.; Stojanović, V. *Clusters of Generation Z and Travel Risks Perception: Constraining vs. Push–Pull Factors In The Emerald Handbook of Destination Recovery in Tourism and Hospitality*; Emerald Publishing Limited: Bingley, UK, 2022; pp. 375–395, eISBN: 978-1-80262-073-3. [CrossRef]
12. Hsu, C.Y.; Chen, M.Y.; Nyaupane, G.P.; Lin, S.H. Measuring sustainable tourism attitude scale (SUS-TAS) in an Eastern island context. *Tour. Manag. Perspect.* **2020**, *33*, 100617. [CrossRef]
13. Ladičorbić, M.M.; Dragin, A.; Stojanović, V.; Lazić, L.; Janjušević, B.; Carmer, A.; Hansen, B. Do We Want to Be Involved? Tourism Promotion and the Industrial Heritage of Germans in Vojvodina. *Sociol. Space/Sociol. Prost.* **2023**, *61*, 79–104. [CrossRef]
14. Jeelani, P.; Shah, S.A.; Dar, S.N.; Rashid, H. Sustainability constructs of mountain tourism development: The evaluation of stakeholders' perception using SUS-TAS. *Environ. Dev. Sustain.* **2023**, *25*, 8299–8317. [CrossRef] [PubMed]
15. Guo, X.; Liu, J.; Chen, Z.; Hong, X.-C. Harmonious Degree of Sound Sources Influencing Visiting Experience in Kulangsu Scenic Area, China. *Forests* **2023**, *14*, 138. [CrossRef]

16. Kalfas, D.G.; Zagkas, D.T.; Dragozi, E.I.; Melfou, K. An approach of landsenses ecology and landsenseology in Greece. *Int. J. Sustain. Dev. World Ecol.* **2021**, *28*, 677–692. [CrossRef]
17. Wang, P.; Li, N.; He, Y.; He, Y. Evaluation of Cultural Ecosystem Service Functions in National Parks from the Perspective of Benefits of Community Residents. *Land* **2022**, *11*, 1566. [CrossRef]
18. Chassang, L.; Hsieh, C.J.; Li, T.N.; Hsieh, C.M. Feasibility Assessment of Stakeholder Benefits in Community-Based Agritourism through University Social Responsibility Practices. *Agriculture* **2024**, *14*, 602. [CrossRef]
19. Pereira, R.; Costa, V.; Gomes, H. Health and Wellness Tourism: An Overview of Thermal Tourism in Portugal. *J. Tour. Sustain. Well-Being* **2023**, *11*, 136–147. [CrossRef]
20. Global Wellness Institute. The Global Wellness Report: Looking Beyond COVID. 2021. Available online: https://globalwellnessinstitute.org/wp-content/uploads/2021/11/GWI-WE-Monitor-2021_final-digital.pdf (accessed on 1 September 2024).
21. Gonçalves, E.C.; da Guerra, R.J.C. O turismo de saúde e bem-estar como fator de desenvolvimento local: Uma análise à oferta termal portuguesa. PASOS. *Rev. Tur. Patrim. Cult.* **2019**, *17*, 453–472. [CrossRef]
22. Bacsi, Z.; Kovács, E.; Lőke, Z. Spa successes and challenges in Transdanubia, Hungary–Results of a survey in three spa towns. *J. Deturope* **2012**, *4*, 27–47. Available online: https://deturope.eu/pdfs/det/2012/01/03.pdf (accessed on 1 September 2024). [CrossRef]
23. Cristian-Constantin, D.; Radu-Daniel, P.; Daniel, P.; Georgiana, C.L.; Igor, S. The role of SPA tourism in the development of local economies from Romania. *Procedia Econ. Financ.* **2015**, *23*, 1573–1577. [CrossRef]
24. Mueller, H.; Kaufmann, E.L. Wellness tourism: Market analysis of a special health tourism segment and implications for the hotel industry. *J. Vacat. Mark.* **2001**, *7*, 5–17. [CrossRef]
25. Maltsev, A.; Fechina, A. Global Health Services Market in the New Economic Conditions. In Proceedings of the 2nd International Scientific Conference on New Industrialization: Global, National, Regional Dimension (SICNI 2018), Ekaterinburg, Russia, 12 April–12 May 2018; pp. 515–519. [CrossRef]
26. Peptenatu, D.; Draghici, C.; Merciu, C. Characteristics of entrepreneurial profile in some emergent territorial structures in Romania. *Actual Probl. Econ.* **2012**, *12*, 448–458.
27. Suess, C.; Baloglu, S.; Busser, J.A. Perceived impacts of medical tourism development on community wellbeing. *Tour. Manag.* **2018**, *69*, 232–245. [CrossRef]
28. PSETOUR. Study of Spa Locations in Vojvodina. The Provincial Secretariat for Economy and Tourism, Novi Sad. 2019. Available online: https://www.pressreader.com/serbia/blic/20190320/281844349970436 (accessed on 1 September 2024).
29. Hoang, T.T.H.; Van Rompaey, A.; Meyfroidt, P.; Govers, G.; Vu, K.C.; Nguyen, A.T.; Hens, L.; Vanacker, V. Impact of tourism development on the local livelihoods and land cover change in the Northern Vietnamese highlands. *Environ. Dev. Sustain.* **2020**, *22*, 1371–1395. [CrossRef]
30. Deng, A.; Lu, J.; Zhao, Z. Rural destination revitalization in China: Applying evolutionary economic geography in tourism governance. *Asia Pacic. J. Tour. Res.* **2020**, *26*, 215–230. [CrossRef]
31. Esfandiari, H.; Choobchian, S. Designing a Wellness-Based Tourism Model for Sustainable Rural Development. 2020. Available online: https://europepmc.org/article/ppr/ppr205797 (accessed on 1 September 2024).
32. Wang, K.; Xu, H.; Huang, L. Wellness tourism and spatial stigma: A case study of Bama, China. *Tour. Manag.* **2020**, *78*, 104039. [CrossRef]
33. Sharpley, R. Host perceptions of tourism: A review of the research. *Tour. Manag.* **2014**, *42*, 37–49. [CrossRef]
34. Tuan, Y.-F. *Space and Place: The Perspective of Experience*; University of Minnesota Press: Minneapolis, MN, USA, 1977.
35. Cheng, C.K. Understanding Visual Preferences for Landscapes: An Examination of the Relationship Between Aesthetics and Emotional Bonding. Ph.D. Dissertation, Texas A & M University, College Station, TX, USA, 2010.
36. Raymond, C.; Brown, G. A spatial method for assessing resident and visitor attitudes towards tourism growth and development. *J. Sustain. Tour.* **2007**, *15*, 520–540. [CrossRef]
37. Lane, B.; Kastenholz, E. Rural tourism: The evolution of practice and research approaches–towards a new generation concept? *J. Sustain. Tour.* **2015**, *23*, 1133–1156. [CrossRef]
38. Ivasciuc, I.S.; Ispas, A. Exploring the motivations, abilities and opportunities of young entrepreneurs to engage in sustainable tourism business in the mountain area. *Sustainability* **2023**, *15*, 1956. [CrossRef]
39. Sirakaya-Turk, E.; Ekinci, Y.; Kaya, A.G. An examination of the validity of SUS-TAS in cross-cultures. *J. Travel Res.* **2008**, *46*, 414–421. [CrossRef]
40. Aprilia, A.; Aini, E.K.; Kumalasari, K.P.; Fajar, Y. The impact of residents attitude toward sustainable tourism development using SUS-TAS: Mediating role of tourism involvement. *Agrisocionomics J. Sos. Ekon. Pertan.* **2024**, *8*, 514–527. [CrossRef]
41. Akis, S.; Peristianis, N.; Warner, J. Residents' attitudes to tourism development: The case of Cyprus. *Tour. Manag.* **1996**, *17*, 481–494. [CrossRef]
42. Gursoy, D.; Chi, C.G.; Dyer, P. Locals' attitudes toward mass and alternative tourism: The case of Sunshine Coast, Australia. *J. Travel Res.* **2010**, *49*, 381–394. [CrossRef]
43. Wang, S.; Chen, S.; Xu, H. Resident attitudes towards dark tourism, a perspective of place-based identity motives. *Curr. Issues Tour.* **2019**, *22*, 1601–1616. [CrossRef]

44. Głąbiński, Z.; Duda, T. The local community perception of tourism development. The case study of Gryfino County-Western Pomerania, Poland. Bulletin of Geography. *Socio-Econ. Ser.* **2017**, *37*, 7–23. [CrossRef]
45. Ribeiro, M.A.; Pinto, P.; Silva, J.A.; Woosnam, K.M. Examining the predictive validity of SUS-TAS with maximum parsimony in developing island countries. *J. Sustain. Tour.* **2018**, *26*, 379–398. [CrossRef]
46. Megeirhi, H.A.; Woosnam, K.M.; Ribeiro, M.A.; Ramkissoon, H.; Denley, T.J. Employing a value-belief-norm framework to gauge Carthage residents' intentions to support sustainable cultural heritage tourism. *J. Sustain. Tour.* **2020**, *28*, 1351–1370. [CrossRef]
47. Rathnayake, C.V.; Darshi, G.A.N. An application of sustainable tourism attitude scale (sus-tas) in three coastal tourist destinations in the southern province of Sri Lanka. In Proceedings of the Sixth International Conference on Business Management, Faculty of Management Studies and Commerce, Guangzhou, China, 26–28 August 2022; University of Sri Jayewardenepura: Nugegoda, Sri Lanka, 2022; pp. 161–170.
48. Tomić, N.; Košić, K. Developing the Spa Assessment Model (SAM) and its application on the Kopaonik-Jastrebac spa zone (Serbia). *Tour. Manag. Perspect.* **2020**, *36*, 100753. [CrossRef]
49. Choi, H.S.C.; Sirakaya, E. Measuring residents' attitude toward sustainable tourism: Development of sustainable tourism attitude scale. *J. Educ. Technol. Syst.* **2005**, *43*, 111–121. [CrossRef]
50. Košić, K.; Pivac, T.; Romelić, J.; Lazić, L.; Stojanović, V. Characteristics of thermal–mineral waters in Backa region (Vojvodina) and their exploitation in spa tourism. *Renew. Sustain. Energy Rev.* **2011**, *15*, 801–807. [CrossRef]
51. Obradović, S.; Stojanović, V. Measuring residents' attitude toward sustainable tourism development: A case study of the Gradac River gorge, Valjevo (Serbia). *Tour. Recreat. Res.* **2022**, *47*, 499–511. [CrossRef]
52. Obradović, S.; Stojanović, V.; Lukić, T. Geotourism and Local Communities: Measuring Residents' Attitudes Toward Sustainable Tourism in the Fruška Gora National Park. *Geoheritage* **2023**, *15*, 90. [CrossRef]
53. Dragin, A.; Mijatov Ladičorbić, M.; Stojanović, V.; Lazić, L.; Komnenić, B.; Damnjanović, J.; Carmer, A.; Hansen, B. Evaluating resident attitudes on German industrial heritage and tourism using the sustainable tourism attitude scale in Vojvodina, Serbia. *J. Herit. Tour.* **2023**, *18*, 864–884. [CrossRef]
54. Ivkov, M.; Mijatov Ladičorbić, M.; Milošević, S.; Dimić, N.; Đoković, F.; Bučić, A.; Nićin, M. Evaluation of stakeholders' attitudes towards sustainable development of congress tourism in Serbia. In Proceedings of the 6th International Scientific Conference Trends in the Development of Tourism and Hospitality, Kotor, Montenegro, 8–9 September 2023. [CrossRef]
55. Chrontsiou, M. *Analyzing the Role of Wellness Tourism in Promoting Economic, Social, and Environmental Sustainability in Municipality of Thermi, Greece*; International Helenic University: Nea Moudania, Greece, 2024.
56. Dragin, A.S.; Mijatov, M.B.; Majstorović, N.; Janičić, B.; Korovljev, D. COVID-19 Pandemic and Young Tourists' Travel Risk Perceptions. In *COVID-19 and a World of Ad Hoc Geographies*; Brunn, S.D., Gilbreath, D., Eds.; A chapter No. 95; Springer: Cham, Switzerland; New York, NY, USA, 2022; Volume 3, pp. 1755–1775, eISBN: 978-3-030-94350-9. [CrossRef]
57. Dragin, A.S.; Mijatov, M.B.; Munitlak Ivanović, O.; Jovičić Vuković, A.; Ivkov Džigurski, A.; Košić, K.; Nedeljković Knežević, M.; Tomić, S.; Stankov, U.; Vujičić, M.D.; et al. Entrepreneurial Intention of Students (Managers in Training): Personal and Family Characteristics. *Sustainability* **2022**, *14*, 7345. [CrossRef]
58. Mijatov, M.B.; Dragin, A.S.; Majstorović, N.; Janičić, B.; Zadel, Z. The impacts of the COVID-19 on youth tourism and hospitality managers in-training. In *COVID-19 and a World of Ad Hoc Geographies*; A chapter No. 102; Brunn, S.D., Gilbreath, D., Eds.; Springer: Cham, Switzerland; New York, NY, USA, 2022; Volume 3, pp. 1879–1895, eISBN: 978-3-030-94350-9; ISBN 978-3-030-94350-9. [CrossRef]
59. Eurostat. Available online: https://ec.europa.eu/eurostat/web/rural-development/methodology (accessed on 3 August 2024).
60. Zheng, T.; Yan, Y.; Lu, H.; Pan, Q.; Zhu, J.; Wang, C.; Zhang, W.; Rong, Y.; Zhan, Y. Visitors' perception based on five physical senses on ecosystem services of urban parks from the perspective of landsenses ecology. *Int. J. Sustain. Dev. World Ecol.* **2020**, *27*, 214–223. [CrossRef]
61. Tešin, A.; Dragin, A.S.; Mijatov Ladičorbić, M.; Jovanović, T.; Zadel, Z.; Surla, T.; Košić, K.; Amezcua-Ogáyar, J.M.; Calahorro-López, A.; Kuzman, B.; et al. Quality of Life and Attachments to Rural Settlements: The Basis for Regeneration and Socio-Economic Sustainability. *Land* **2024**, *13*, 1364. [CrossRef]
62. Han, L.; Shi, L.; Yang, F.; Xiang, X.; Gao, L. Method for the evaluation of residents' perceptions of their community based on landsenses ecology. *J. Clean. Prod.* **2021**, *281*, 124048. [CrossRef]
63. Harun, R.; Chiciudean, G.O.; Sirwan, K.; Arion, F.H.; Muresan, I.C. Attitudes and perceptions of the local community towards sustainable tourism development in Kurdistan regional government, Iraq. *Sustainability* **2018**, *10*, 2991. [CrossRef]
64. Alshboul, K. Assessing Local Community Involvement in Tourism Development Around a Proposed World Heritage Site in Jerash, Jordan. Ph.D. Thesis, University of Waterloo, Waterloo, ON, Canada, 2016.
65. Carpenter, M.; Miller, C. Revitalizing community development in theory and practice? *Community Dev. J.* **2011**, *46*, i1–i6. [CrossRef]
66. Kim, S.; Park, E.; Phandanouvong, T. Barriers to local residents' participation in community-based tourism: Lessons from Houay Kaeng Village in Laos. *SHS Web Conf.* **2014**, *12*, 01045. [CrossRef]
67. Petrović, M.D.; Blešić, I.; Vujko, A.; Gajić, T. The role of agritourism's impact on the local community in a transitional society: A report from Serbia. *Transylv. Rev. Adm. Sci.* **2017**, *13*, 146–163. [CrossRef]

68. Campón-Cerro, A.M.; Folgado-Fernández, J.A.; Hernández-Mogollón, J.M. Rural destination development based on olive oil tourism: The impact of residents' community attachment and quality of life on their support for tourism development. *Sustainability* **2017**, *9*, 1624. [CrossRef]
69. McGehee, N.G.; Andereck, K.L. Factors predicting rural residents' support of tourism. *J. Travel Res.* **2004**, *43*, 131–140. [CrossRef]

land MDPI

Article

Influence of Livelihood Capitals on Landscape Service Cognition and Behavioral Intentions in Rural Heritage Sites

Shiying Li [1], Yaqi Cheng [1], Jiayu Cai [1,2] and Xuewei Zhang [1,2,*]

1 School of Architecture and Urban-Rural Planning, Fuzhou University, Fuzhou 350108, China; 221520018@fzu.edu.cn (S.L.); 231520002@fzu.edu.cn (Y.C.); 221520001@fzu.edu.cn (J.C.)
2 Fujian Key Laboratory of Digital Technology for Territorial Space Analysis and Simulation, Fuzhou 350108, China
* Correspondence: xuewei.zhang@fzu.edu.cn

Abstract: Farmers' livelihoods are critical for global sustainable development and the conservation and transmission of rural heritage. However, neglecting farmers' livelihoods increases the risks to living heritage conservation. Therefore, it is essential to explore the impact of livelihood capital on farmers' landscape services cognition and their behavioral intentions. Based on the Sustainable Livelihoods Approach, this study examines the rural areas of Mulanbei irrigation district, which was recognized as a World Heritage Irrigation Structure in 2014, and uses a structural equation model to investigate the relationships between farmers' livelihood capital, landscape services cognition and behavioral intentions. The study's key findings include the following: (1) The levels of human capital (0.541), social capital (0.671), and cultural capital (0.645) are relatively high, while the levels of natural, physical, and financial capital are comparatively low. (2) There are significant differences in landscape service cognition and behavioral intentions among farmers of different livelihood strategies, with diversified livelihood farmers demonstrating the highest levels of both cognition and intentions, while subsidy-dependent farmers show the lowest levels. (3) Natural, cultural and financial capital play a crucial role in influencing farmers' landscape services cognition and their behavioral intentions. Landscape service cognition mediates the relationship between livelihood capital and behavioral intentions. (4) To increase farmers' willingness to protect and promote rural heritage, efforts should focus on enhancing natural, physical, and financial capital while fostering cultural capital to promote advocacy.

Keywords: sustainable livelihoods; human settlement; world heritage irrigation structures; livelihood capital; landscape service cognition; behavioral intentions

1. Introduction

A livelihood comprises the assets (including both material and social resources), capabilities, and activities that an individual or household possesses for making a living and improving long-term well-being [1]. Farmers' livelihood conditions directly affect their quality of life and long-term welfare. Rural heritage is closely related to farmers' settlement behaviors and their agricultural and pastoral activities [2], reflecting the positive interaction between farmers' livelihoods and rural resource management. It serves as a valuable model for sustainable human–environment relationships. Recently, due to the influences of urbanization, industrialization, and commercialization, farmers' income sources have become more diversified, resulting in changes in household income, employment prospects, and educational and training opportunities. Simultaneously, farmers' perceptions and attitudes toward traditional resource management practices have undergone significant shifts. On the one hand, consensus holds that rural heritage contributes to diversified livelihoods [3,4], promoting the development of part-time farming and increasing income levels [5–7]. On the other hand, the single-directional transformation of farmers' livelihoods in the context of heritage tourism may reduce livelihood resilience and stability and even damage the authenticity and vitality of heritage due to the excessive pursuit of economic benefits [8]. Nevertheless, the long historical evolution of rural heritage has deeply ingrained a sense of connection between local farmers and their heritage. However, in developing countries, government authorities often lead heritage conservation efforts. When the responsibility for heritage protection and utilization is divided, the intended benefits for local farmers may not be fully realized, and farmers may not recognize the social and economic advantages. Recent studies emphasize that farmers' active involvement is crucial for effective heritage conservation, as they are the smallest decision-making units in these efforts [9,10]. Furthermore, farmers' behavioral intentions are shaped by their livelihood characteristics, with a tendency to optimize their livelihood capital [11,12]. Therefore, understanding the challenges farmers face in selecting livelihood strategies is essential for achieving the dual goals of heritage conservation and community development.

Sustainable livelihoods can cope with and recover from stress and shocks, maintain or enhance its capabilities and assets, and provide sustainable livelihood opportunities for the next generation while contributing net benefits to other livelihoods at the local and global levels and in the short and long term [1,13]. In 1997, the UK Department for International Development (DFID) issued a white paper, 'Eliminating Global Poverty: The Challenge for the 21st Century', which explicitly proposed promoting human development and environmental protection by creating sustainable livelihoods for the poor. Since then, extensive theoretical and practical exploration has emerged concerning rural livelihoods [14], such as sustainable livelihood analysis [15], and livelihoods in tourism destinations [16]. These efforts have led to the development of livelihood analysis frameworks that emphasize the interrelationships such as "livelihood context–livelihood capital–institutions and processes–livelihood strategies–livelihood outcomes." Among these theoretical frameworks, the Sustainable Livelihoods Approach (SLA) developed by the DFID has been widely adopted, dividing livelihood capital into five fundamental types: human, natural, physical, financial, and social [15]. This approach was later expanded to include cultural and tourism capital, allowing for a more comprehensive characterization of household features and the diagnosis of livelihood development bottlenecks. Sustainable livelihoods have emerged as an important approach for addressing gaps in traditional ecosystem services and resource management, particularly in terms of recognizing social choices and stakeholder perspectives [17]. This approach focuses on exploring win–win solutions that optimize landscape functions and improve well-being by considering the diverse needs of farmers and their livelihoods [18].The SLA has been acknowledged as a vital tool for examining rural issues of poverty alleviation and development, effectively linking environmental characteristics, social demands, and individual decision-making. This not only helps us to understand how micro-level forces shape landscape evolution [19,20] but also ensures that landscape decision-making is aligned with the interests of relevant stakeholders [21,22], providing a theoretical foundation for investigating the relationship between rural heritage landscape provision and farmers' behavioral decisions.

World Heritage Irrigation Structures (WHISs) represent a prime example of rural productive landscape heritage. Over time, irrigation districts have evolved into an agricultural-based landscape structure and a rural society centered on water management, making significant contributions to increasing food production, improving farmers' livelihoods, and promoting rural prosperity. These WHISs, consisting of heritage sites and their surroundings, are always vast and complex. Their extensive surface canal systems and continued functioning heavily depend on the spontaneous maintenance efforts of local farmers. Despite the strong attachment to and sense of identity that farmers have with these heritage sites, livelihood pressures often lead to behaviors conflicting with the goals of sustainable heritage development. Examples include encroaching on water bodies for expanding residential and industrial areas, overusing agricultural fertilizers, or opting out of local cultural activities. Previous studies have examined how external factors such as individual characteristics, household structures, income levels, social networks, and agricultural policies influence farmers' behavioral intentions [23–25].

As the autonomous initiative of farmers continues to strengthen, research on internal cognitive motivations and the dynamic processes driving behavior change has developed progressively. Li et al. examined how farmers' perceptions impact their willingness to engage in green agricultural practices, identifying specific behavioral factors [26,27]. Building upon this, Wang et al. investigated the mediating role of livelihood capital and value perception in influencing pro-environmental behavior, adding depth to the understanding of these motivations [28]. More recently, Lyu et al. expanded this work by exploring the relationship between farmers' cognition, intentions, and behaviors in the context of sustainable land intensification in Shandong Province, offering a comprehensive view of these dynamics [23,29]. Despite these advances, there is still limited research on how farmers perceive the value of landscape services and how this perception influences their engagement and acceptance of conservation measures. This might lead to a mismatch between policy recommendations and the actual needs and expectations of farmers. As research on rural heritage conservation increasingly shifts from a purely physical focus to exploring the interaction between farmers' cognition and their behavioral intentions [4,30], an important question arises: how can enhancing farmers' awareness of landscape services guide their behavioral intentions to achieve optimal livelihood outcomes? This question has become a central focus in both academic and practical discussions.

This study aims to examine farmers' cognition of the landscape services provided by WHISs and their associated behavioral intentions. It seeks to uncover the key factors influencing farmers' behavioral intentions and explore how these intentions differ among farmers with various livelihood strategies. The results will offer valuable insights for developing tailored policies that balance heritage conservation and utilization. Specifically, this study addresses three main questions:

(1) How do farmers with different livelihood strategies differ in their cognition of landscape services and behavioral intentions?
(2) What are the specific pathways through which different types of livelihood capital influence landscape service cognition and behavioral intentions?
(3) Does landscape service cognition mediate the relationship between livelihood capital and farmers' behavioral intentions?

2. Research Framework and Research Hypotheses

2.1. Research Framework

The core idea of livelihood studies is to affirm that small-scale farmers possess the wisdom and potential to make effective behavioral decisions. Livelihood capital serves as a key foundation for these decisions, providing a multidimensional representation of farmers' assets and capabilities. For instance, the accumulation of human capital may enable farmers to participate more actively in conservation and promotional activities, while abundant natural capital can enhance their awareness of environmental protection. From the perspective of the SLA, farmers can adopt livelihood strategies in response to changes in the external environment, which subsequently influence both the characteristics and the condition of their livelihood capital [31]. However, recent studies suggest that farmers' livelihood strategies are not always entirely rational, especially in the context of rural tourism development. Changes in livelihood strategies may lead to disturbances in the socio-ecological system, resulting in short-term increases in livelihood capital but heightened vulnerability [32]. Moreover, research has found that when faced with decisions regarding resource development and conservation, farmers tend to opt for familiar approaches rather than experimenting with innovative methods or new policy tools [33]. This cognitive bias can affect farmers' long-term judgments about the value of ecosystems, leading to irrational decisions.

Cognition and Behavior Theory (CBT) posits that cognitive processes play a crucial role in behavioral modification, emphasizing that behavioral decisions stem from an individual's behavioral intentions (BI). BI, in turn, is shaped by cognitive perceptions formed during the process of receiving and evaluating information [34]. Numerous studies have demonstrated that CBT significantly enhances the explanatory power of behavioral intentions research, particularly in fields such as health psychology and social behavior [35]. Its applicability has also been increasingly recognized in environmental behavior studies [36]. Some scholars have applied CBT to explore farmers' behavioral intentions, finding that farmers' attachment to their homeland and sense of belonging can increase their willingness to participate in rural environmental protection [37–39]. However, its connection with farmers' specific livelihood needs remains under explored. Therefore, this study integrates CBT with SLA to comprehensively analyze the mechanisms underlying farmers' behavioral intentions.

Specifically, rural heritage, as a form of living heritage, relies on the important and continuous landscape services it provides to local farmers. Over centuries, WHISs have offered key landscape services to farmers in the irrigation districts, such as storm-water management, climate regulation, food supply, and biodiversity conservation. With socio-economic development and changes in resource utilization, the heritage's primary functions have gradually shifted from being predominantly ecological and productive to a combination of ecological and social functions [40]. Farmers' behavioral intentions are generally aligned with the goal of increasing livelihood capital and harmonizing with livelihood strategies. At the same time, their perception of the landscape services provided by heritage also influences their behavioral intentions. Some farmers may even be willing to sacrifice personal interests to contribute to the collective goals of the local community. This perception is not only linked to the objective landscape services provided by the heritage but is also influenced by farmers' specific livelihood strategies.

To capture these dynamics, this study incorporates CBT into the SLA, constructing an analytical model that encompasses livelihood capital, landscape service cognition, and behavioral intentions. As shown in Figure 1, livelihood capital and farmers' cognition directly influence behavioral intentions and play a role in long-term livelihood strategies. Additionally, considering that rural heritage sites are typically shaped by unique cultural characteristics and socio-ecological relationships [41], this study introduces "cultural capital" as a distinct component of livelihood capital, alongside the traditional five types of livelihood capital.

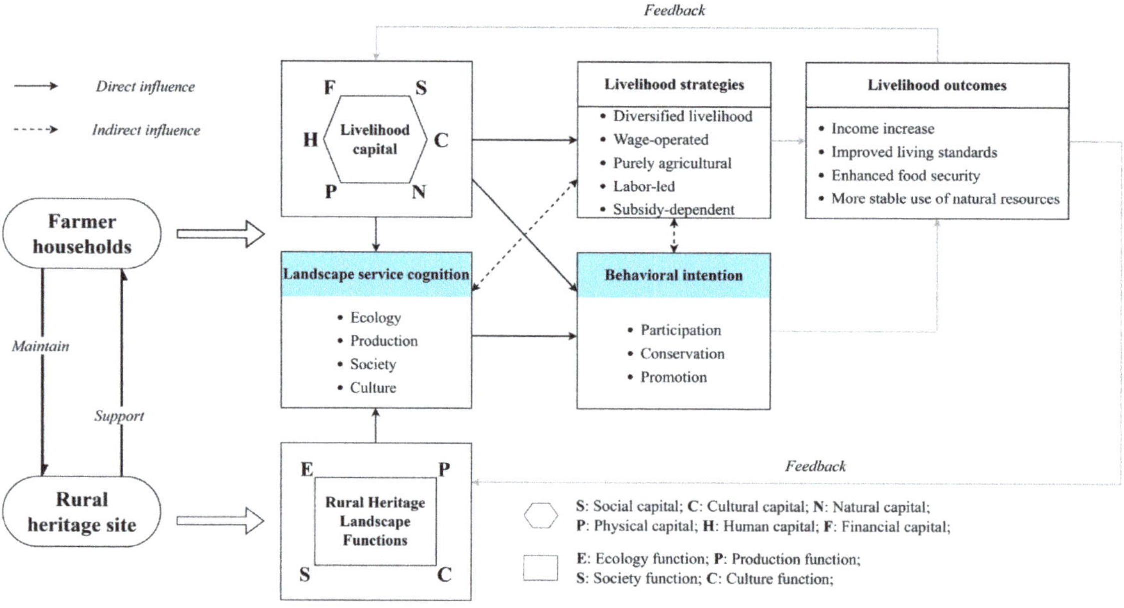

Figure 1. The research framework integrates CBT into the SLA in rural heritage sites.

2.2. Research Hypotheses

Farmers' willingness to engage in rural heritage protection and utilization reflects their attitudes and decision-making processes regarding environmental and resource management. Based on the proposed research framework, the relationships among farmers' livelihood capital, landscape service cognition, and their willingness to engage in heritage protection and utilization behaviors are articulated as follows:

Hypothesis 1. *Different types of livelihood capital have a significant positive impact on the willingness to participate, conserve and promote.*

Hypothesis 2. *Different types of livelihood capital have a significant positive impact on landscape service cognition.*

Hypothesis 3. *Landscape service cognition has a significant positive impact on the willingness to participate, to conserve and to promote.*

Hypothesis 4. *Landscape service cognition mediates the effects of different types of livelihood capital on willingness to participate, conserve, and promote.*

3. Materials and Methods

3.1. Study Area

The Mulanbei Water Conservancy Project (MWCP), located in Putian, Fujian Province, China, started construction in 1064 BC and was completed in 1083 BC during the Northern Song Dynasty, with three rounds of site re-selections and reconstructions. Having been in operation for more than 930 years, it has largely preserved both the original position and form of its engineering structures from the historical period. It is recognized as the largest dam diversion project with saltwater rejection and freshwater conservation functions in ancient Fujian Province. In 2014, it was listed as one of the first World Heritage Irrigation Structures (WHISs). The main engineering heritages include three parts: the weir, the canal system, and the embankment project. The weir is primarily represented by the Mulanbei dam, which spans a length of 219.13 m. The total length of the canal system is 309.5 km, while the embankment project extends for 87.48 km. According to statistics from the International Commission on Irrigation and Drainage (ICID), the effective irrigated area covers 91.33 km^2 and benefits over 500,000 people. The MWCP is acknowledged as a unique plain granary and an important water conservation area in the southeast coast of China, as well as one of the regions experiencing the fastest coastal urbanization development [42].

In response to increasing urban development pressures and the lack of guidance for internal village construction, Putian City initiated four rounds of "Ecological Green Center Planning" from 2008 to 2023. Land use planning drawings can be found in Appendix A. This initiative aims to protect areas with a concentrated distribution of cultural and natural resources within the Mulanbei irrigation district. The master planning, confirmed in September 2023, delineates a core protected area of 65 km^2, encompassing 61 villages and 3 non-village areas (including one marine area and 2 state-owned or town-owned tidal flats). Among these, 10 villages possess distinctive protective value, 21 villages exhibit industrial characteristics, and 4 are recognized as provincial historical and cultural villages. Considering factors such as village location, industrial characteristics, landscape features, and land use, 7 villages with the highest protective value were selected as research samples. These include 3 villages with industrial characteristics—Daili Village (a Taobao village), Qibu Village (shoe-making industry), and Chenqiao Village (furniture industry); one provincial traditional village—Dongyang Village; one key cultural tourism development village—Wujiang Village; and 2 provincial water conservancy scenic spots—Jiangdong Village and Dongjia Village. This is shown in Figure 2.

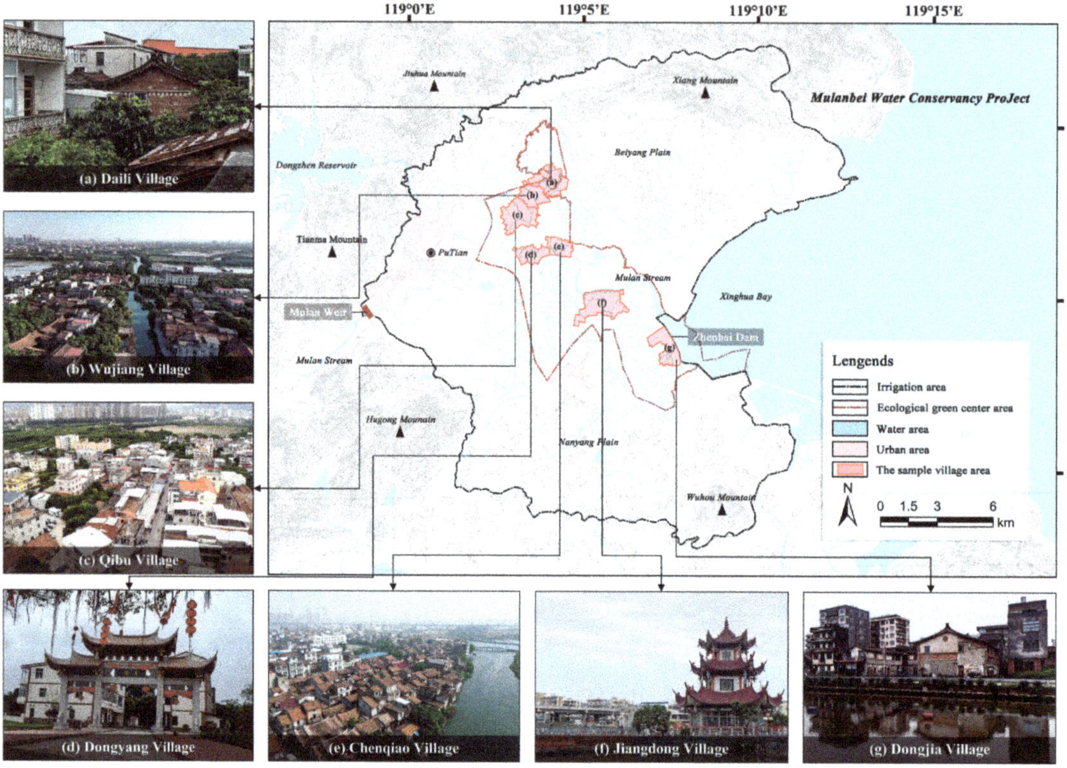

Figure 2. Study area. (Photo source: All photos are taken by the author).

3.2. Questionnaire Design

The questionnaire content encompasses four key aspects:

1. **Basic Socioeconomic Characteristics of the Respondents:** This consisted of education level, household size, etc.
2. **Livelihood Capital Status of the Respondents:** This aspect examines various types of capital, including human capital, natural capital, physical capital, financial capital, social capital, and cultural capital.
3. **Respondents' Cognition of Landscape Services in the Area:** Respondents are asked to evaluate their awareness and understanding of landscape services based on their own perceptions.
4. **Behavioral Willingness of the Respondents:** This consisted of willingness to participate, willingness to protect, and willingness to promote.

Aspects 1 and 4 in the list above included respondents' scores according to a Likert scale.

3.3. Variable Selection and Measurement

This study builds upon the research of Ren et al. [43] and, in conjunction with the framework established earlier, designs a measurement scale for farmers' livelihood capital, landscape service cognition, and behavioral willingness. This scale encompasses three categories of variables:

1. **Independent Variable (IV):** Farmers' livelihood capital, which includes 12 indicators across six dimensions: human capital, natural capital, physical capital, financial capital, social capital, and cultural capital.

2. **Mediating Variable (MV):** Farmers' cognition of landscape services, comprising 11 measurement indicators within four dimensions: ecology, production, society, and landscape culture.
3. **Dependent Variable (DV):** Farmers' behavioral intentions, which include three dimensions: willingness to participate, willingness to protect, and willingness to publicize, totaling 11 items.

The measurement of livelihood capitals and the specific variable settings are presented in Tables 1 and 2.

Table 1. Indicators and quantification of livelihood capital.

Livelihood Capital	Indicator	Magnitude Definition
Human capital	Education level (A1)	Illiteracy = 1; Elementary school and below = 2; Junior high school = 3; High school/technical secondary school = 4; Junior college/higher vocational college and above = 5
	Proportion of household labor force (A2)	0~≤1/5 = 1; 1/5~≤2/5 = 2; 2/5~≤3/5 = 3; 3/5~≤4/5 = 4; 4/5~≤1 = 5
Natural capital	Household cropland area (A3)	0~≤1 mu = 1; 1~≤2 mu = 2; 2~≤3 mu = 3; 3~≤4 mu = 4; >4 mu = 5
	Water and fertilizer conditions of cropland (A4)	Very poor = 1; Poor = 2; Neutral = 3; Good = 4; Very good = 5
Physical capital	Types of household appliances (A5)	0 ~ 2 items = 1; 3 ~ 4 items = 2; 5 ~ 6 items = 3; 7 ~ 8 items = 4; More than 8 items = 5
	Types of residential housing (A6)	Old house = 1; Renovated old house = 2; Refurbished house = 3; Newly built house = 4
Financial capital	Household savings status (A7)	30,000~60,000 yuan = 1; 60,000 to 100,000 yuan = 2; 100,000 to 300,000 yuan = 3; 300,000 to 500,000 yuan = 4; More than 500,000 yuan = 5
	The difficulty of lending money to others (A8)	Very difficult = 1; Difficult = 2; Average = 3; Easy = 4; Very easy = 5
Social capital	The closeness with relatives and friends (A9)	Never interact = 1; Rarely interact = 2; Occasionally interact = 3; Fairly often interact = 4; Frequently interact = 5
	Willingness to participate in village activities (A10)	Never participate = 1; Occasionally participate passively = 2; Occasionally participate actively = 3; Frequently participate = 4; Lead participation = 5
Cultural capital	The level of understanding of folklore (A11)	Not familiar at all = 1; Slightly unfamiliar = 2; Neutral = 3; Somewhat familiar = 4; Very familiar = 5
	The degree of recognition of rural cultural values (A12)	Do not approve at all = 1; Somewhat disapprove = 2; Neutral = 3; Somewhat approve = 4; Strongly approve = 5

Table 2. Variable setup and assignment.

Variable Setup			Magnitude Definition
Landscape service cognition	Climate regulation	The climate here is comfortable and pleasant (B1)	1 = Strongly disagree 2 = Disagree 3 = Neutral 4 = Agree 5 = Strongly agree
	Storm-water management	There is no risk of flooding here (B2)	
	Freshwater supply	The water quality here is excellent (B3)	
	Habitat maintenance	There is a great diversity of species here (B4)	
	Food supply	There is a large production of grains, vegetables, and fruits here (B5)	
	Residential support	There are many people living here (B6)	
	Employment security	There are many ways to earn money here (B7)	
	Transportation	The transportation here is very convenient (B8)	
	Cultural value	The culture here makes me feel proud (B9)	
	Landscape aesthetics	The scenery here is beautiful (B10)	
	Recreation and leisure	This place meets the needs for recreation and leisure (B11)	

Table 2. *Cont.*

		Variable Setup	Magnitude Definition
Behavioral Intentions	Participation intention	Support eco-agriculture and leisure tourism for rural heritage conservation. (C1) Invest time/funds in eco-agriculture. (C2) Invest time/funds in handicrafts and processing. (C3) Invest time/funds in tourism. (C4)	1 = Strongly unwilling 2 = Unwilling 3 = Neutral 4 = Willing 5 = Strongly willing
	Conservation intention	The overall willingness to protect the Mulanbei rural heritage. (C5) Actively engage in environmental protection. (C6) Accept industrial land transition subsidies to protect rural heritage. (C7) Embrace farmland transfer benefits for heritage. (C8) Rent and renovate idle homes for preservation. (C9) Donate to protect rural heritage. (C10)	
	Promotion intention	Participate in local cultural promotions. (C11) Encourage others to preserve rural characteristics. (C12)	

3.4. Data Analysis

Considering the arrangements of activities such as employment and farming, and to ensure the representativeness of the surveyed farmers, in this study, we communicated with village leaders, who recommended representative farming households to participate in the questionnaire survey. Random household surveys were then conducted along the main streets of the villages. Given the average number of households (approximately 900) in the sample villages—Chenqiao Village (955 households), Daili Village (356 households), Dongjia Village (1213 households), Dongyang Village (558 households), Jiangdong Village (1371 households), Qibu Village (1182 households), and Wujiang Village (705 households)—around 50 (5%) households were selected from each village for investigation. A total of 371 questionnaires were collected, with 350 valid responses, yielding an effective rate of 94.3%. The distribution of valid questionnaires was as follows: Chenqiao Village (45), Daili Village (59), Dongjia Village (43), Dongyang Village (33), Jiangdong Village (58), Qibu Village (67), and Wujiang Village (45).

Structural equation model (SEM) is a multivariate statistical method designed to analyze complex relationships among multiple indicators. It is suitable for three-dimensional and multi-level analyses, allowing for the quantification of causal relationships between various factors, making it ideal for examining intricate data structures [44]. In practice, the structural equation model has been widely applied in the fields of social sciences, psychology, and management, as well as in other research. In this study, the structural equation model analyzes the path coefficients of latent and observed variables, revealing the influence of farmers' livelihood capital on their cognition of and behavioral intentions toward landscape services. The structural equation model consists of two components: (1) the measurement model, which reflects the relationship between observed and latent variables, and (2) the structural model, which describes the relationships among the latent variables. The equations are as follows.

$$\begin{cases} Y_i = \Lambda_y \eta_i + \varepsilon_i \\ X_i = \Lambda_x \xi_i + \delta_i \end{cases} \tag{1}$$

$$\eta_i = B\eta_i + \Gamma\xi_i + \zeta_i \tag{2}$$

In the equations, Y_i is the endogenous manifest variables; X_i is the exogenous manifest variables; η_i represents the endogenous latent variables; ξ_i is the exogenous latent variable; Λ_y refers to the factor loading matrix for the cognition and behavioral intentions indicators; Λ_x is the factor loading matrix for the livelihood capital indicators; ε_i and δ_i represent variable errors; B and Γ are the coefficient matrices between the endogenous latent variable the exogenous latent variable; and ζ_i denotes the residuals.

4. Results

4.1. Descriptive and Analysis

As shown in Figure 3, 56.3% of respondents were male. The majority of respondents were aged 35 and above, accounting for 89.7%. The education level was generally low, with 72.6% having completed junior high school or below. Most of the respondents were engaged in farming, migrant work, or self-employment (including handicrafts, processing, and commerce), comprising 74.9% of the total.

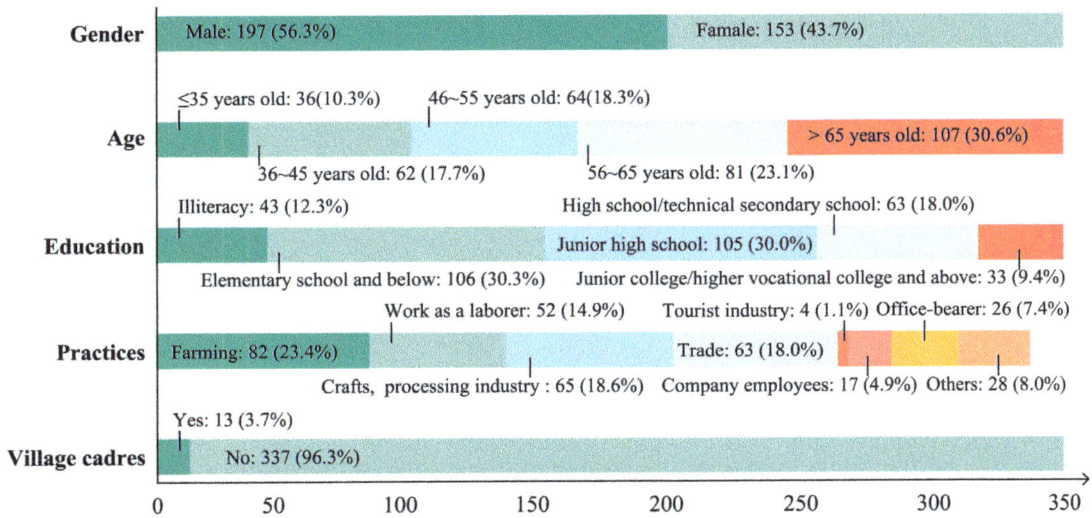

Figure 3. Demographic and sociological characteristics of the surveyed households. (N = 350).

Based on the classification of livelihood strategies in existing studies [14,45], farmers were categorized into five types: diversified livelihood, wage-operated, purely agricultural, labor-led, and subsidy-dependent. As depicted in Table 3, the wage-operated type is the most prevalent, while the traditional purely agricultural type still accounts for more than 20%. A smaller portion of households rely on stable public services or social security, falling into the diversified livelihood and subsidy-dependent categories.

Table 3. Classification of farmer strategies.

Livelihood Strategies	Quantity (Household)	Proportion (%)	Living Mode
Diversified livelihood	39	11.1	Mainly to civil servants, public institutions (including village cadres)
Wage-operated	149	42.6	Enterprise employees, engaging in handicrafts, processing, tourism, commerce, etc.
Purely agricultural	82	23.4	Mainly farming
Labor-led	52	14.9	Mainly for workers
Subsidy-dependent	28	8	Other (mostly retired, unemployed)

Based on the dimensional normalization calculation of the livelihood capital of farmers with different livelihood strategies, the specific results are presented in Figure 4. In general, the relatively high average values of social capital and cultural capital among farmers indicate strong cohesion, resource access, and cultural identity. In contrast, the relatively low levels of natural capital, physical capital, and financial capital suggest weak physical and economic foundations, which may hinder agricultural productivity and economic development.

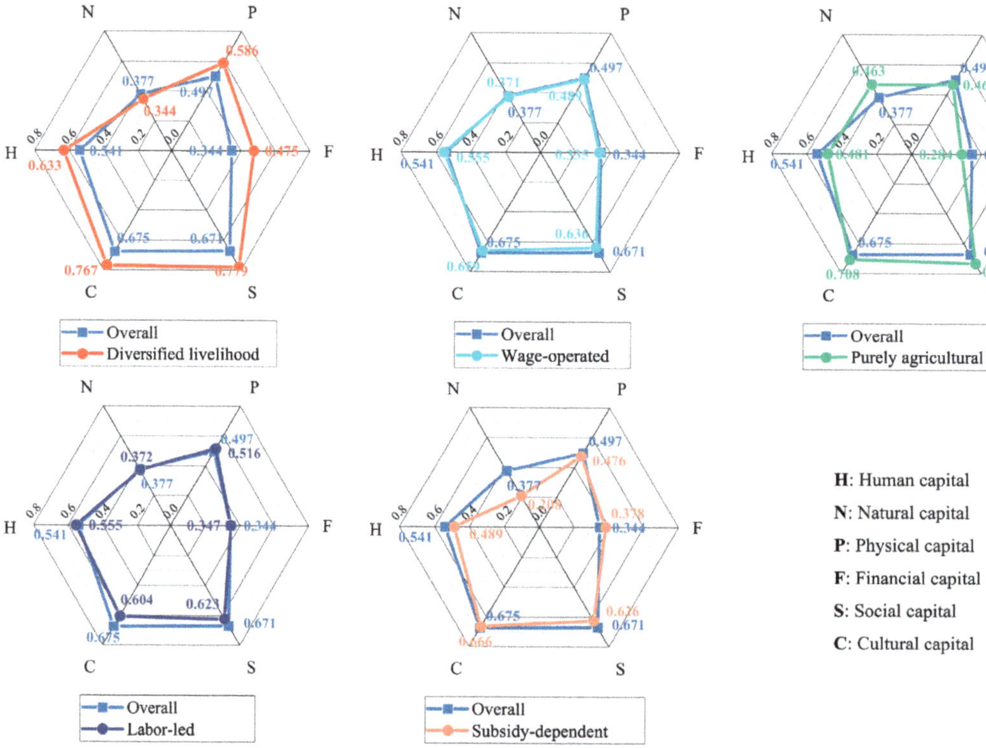

Figure 4. Comparison of average livelihood capital across different strategies of farmers' livelihoods.

Farmers with different livelihood strategies show varied performance across different dimensions of livelihood capital. Compared with the overall average, diversified livelihood farmers perform well across all types of capital, with especially high levels of social capital and cultural capital, reflecting their balanced development capabilities. Wage-operated farmers have livelihood capital levels that are close to the overall average. Purely agricultural farmers excel in natural capital, social capital, and cultural capital. Labor-led farmers exhibit higher levels of human capital and financial capital but lower levels of social capital and cultural capital. However, subsidy-dependent farmers exhibit obviously lower levels of natural capital.

4.2. Differences in Landscape Service Cognition Among Farmers with Different Livelihood Strategies

The difference analysis in Figure 5 shows that the variation in cultural function cognition is relatively significant, while the differences in ecological function cognition are less pronounced. In terms of cultural landscape services, diversified livelihood farmers have a significantly higher recognition of local cultural values compared to wage-operated, subsidy-dependent, and labor-led farmers. Wage-operated farmers show lower awareness of landscape aesthetics and leisure and recreation functions compared to diversified livelihood and purely agricultural farmers. In the dimension of ecological landscape services, only wage-operated farmers exhibit a significantly lower cognition of climate regulation functions compared to diversified livelihood farmers, while the cognitive differences in other ecological functions (such as storm-water management and habitat maintenance) are not significant among farmers of different livelihood strategies. Overall, diversified livelihood farmers demonstrate a higher level of awareness across multiple dimensions of landscape services, particularly in cultural and ecological aspects, making them the group with the most comprehensive and in-depth cognition of landscape services. Additionally, although there are no significant differences in the cognition of transportation (social) and freshwater supply (ecological) among farmers of different livelihood strategies, the overall scores in these aspects are moderately high.

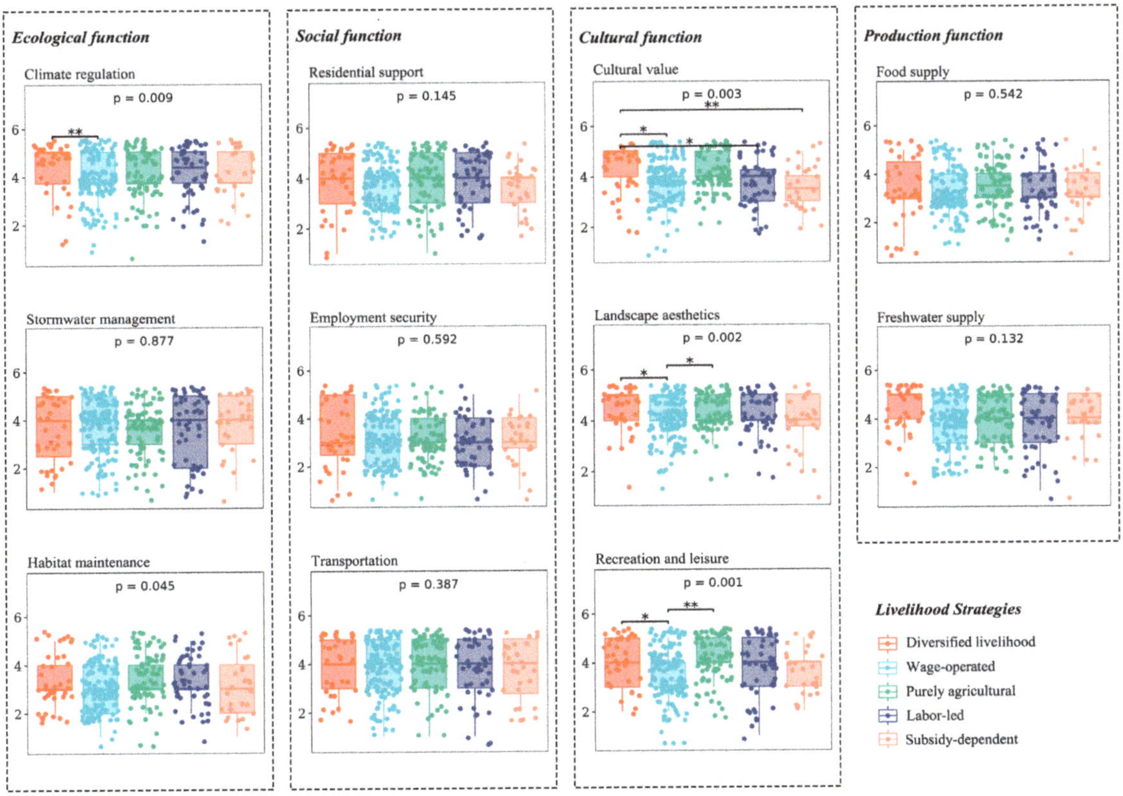

Figure 5. Differences in the cognition of landscape services among farmers with different livelihood strategies. Note: * is $p < 0.05$; ** is $p < 0.01$.

4.3. Differences in the Behavioral Intentions Among Farmers with Different Livelihood Strategies

As shown in Figure 6, farmers of different livelihood strategies exhibit significant differences across three dimensions: participation intention (C1–C4), conservation intention (C5–C10), and promotion intention (C11–C12). In terms of participation intention, diversified livelihood and purely agricultural farmers show the highest overall recognition and support for participating in ecological agriculture, handicrafts, processing industries, and tourism. Purely agricultural farmers, in particular, demonstrate a strong enthusiasm for participating in ecological agricultural production. Regarding protection intention, diversified livelihood farmers express a greater willingness to protect rural heritage sites and the environment. Specifically, their willingness to utilize idle resources and pay fees to protect heritage sites is significantly higher than that of other livelihood strategies. Lastly, for promotion intention, diversified livelihood farmers show a stronger commitment to promoting activities and educating others, reflecting their active involvement in rural protection and development.

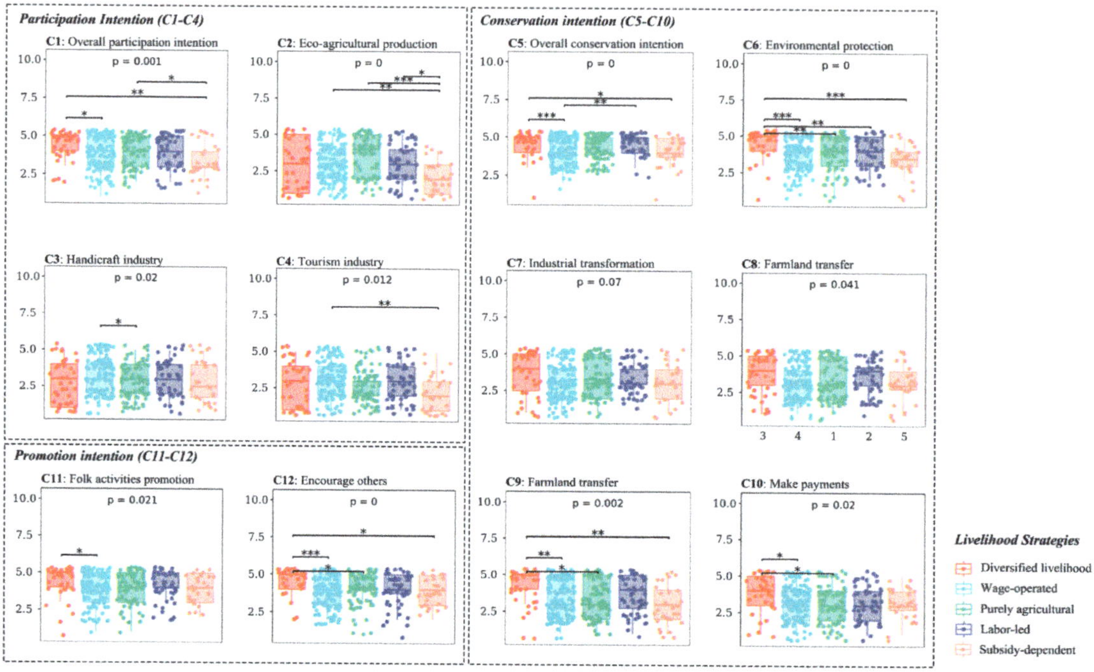

Figure 6. Differences in behavioral intentions among farmers with different livelihood strategies. Note: * is $p < 0.05$; ** is $p < 0.01$.

Overall, diversified livelihood farmers exhibit higher levels of recognition and willingness across all aspects, indicating that they are more proactive in supporting rural protection and development. In contrast, subsidy-dependent farmers display lower enthusiasm for participation and protection but show a relatively higher willingness to engage in the promotion of local cultural traditions.

4.4. Results of Structural Equation Model Regression

The data were analyzed using the statistical software SPSS 27.0. The Cronbach's α coefficient for each latent variable ranged from 0.686 to 0.826, the KMO value was not less than 0.5, and Bartlett's spherical test reached a significance level of 0.01. The results indicate that the sample data have good reliability and validity. The specific results are shown in Table 4. Additionally, the measurement of livelihood capital in this study includes both tangible and intangible dimensions, using context-specific indicators that combine qualitative and quantitative data. Since these factors are not based on a typical Likert scale, traditional reliability tests are not applicable [9]. Instead, the indicators are adapted from well-established frameworks in the literature, carefully selected to align with the study's context and theoretical foundation, ensuring they comprehensively capture the dimensions of livelihood capital [43].

Table 4. Reliability and validity testing of questionnaire across dimensions.

Dimension	Cronbach' s α	KMO
Landscape service cognition	0.773	0.819
Participation intention	0.686	0.648
Conservation intention	0.815	0.780
Promotion intention	0.826	0.500

The structural equation model was constructed using Amos 27.0 software, with parameter estimation performed through the maximum likelihood estimation method. As shown in Table 5, the model was revised based on modification indices. The X^2/df, AGFI, and RMSEA all met the required standards, while the GFI was close to the recommended value. Therefore, the overall model demonstrates a good fit.

Table 5. Reliability and validity testing of questionnaire across dimensions.

Dimension Fit Index	Recommended Value	Fit Value
X^2	The smaller the better	860.531
X^2/df	<3.0	2.502
GFI	>0.9	0.853
AGFI	>0.8	0.814
RMSEA	<0.08	0.066

As shown in Table 6 and Figure 7, regarding the impact of various types of livelihood capital on behavioral intentions, the paths H1d, H1e, H1h, H1i, H1k, H1l, and H1r are significantly established, while the other paths are not significant. Among these, the path coefficients for H1d, H1e, H1k, H1l, and H1r are positive, indicating that natural capital has a significant positive impact on participation and protection intentions; financial capital has a significant positive impact on conservation and promotion intentions; and cultural capital has a significant positive impact on promotion intention. However, the path coefficients for H1h and H1i are negative, suggesting that physical capital has a significant negative impact on conservation and promotion intentions.

Table 6. Assuming results and path coefficient.

Hypothesis	Path Relationship	Estimated Value	S.E.	C.R	*p*-Value	Result
H1a	Human capital → Participation intention	0.090	0.630	1.423	0.155	Not accept
H1b	Human capital → Conservation intention	0.048	0.029	1.626	0.104	Not accept
H1c	Human capital → Promotion intention	0.018	0.057	0.321	0.749	Not accept
H1d	Natural capital → Participation intention	0.128	0.046	2.794	*	Accept
H1e	Natural capital → Conservation intention	0.056	0.022	2.593	**	Accept
H1f	Natural capital → Promotion intention	0.003	0.041	0.063	0.950	Not accept
H1g	Physical capital → Participation intention	−0.002	0.062	−0.033	0.973	Not accept
H1h	Physical capital → Conservation intention	−0.063	0.030	−2.112	*	Accept
H1i	Physical capital → Promotion intention	−0.110	0.057	−2.090	*	Accept
H1j	Financial capital → Participation intention	0.081	0.060	1.342	0.180	Not accept
H1k	Financial capital → Conservation intention	0.112	0.030	3.724	***	Accept
H1l	Financial capital → Promotion intention	0.113	0.055	2.069	*	Accept
H1m	Social capital → Participation intention	0.070	0.058	1.189	0.234	Not accept
H1n	Social capital → Conservation intention	−0.008	0.028	−0.276	0.783	Not accept
H1o	Social capital → Promotion intention	0.059	0.054	1.089	0.276	Not accept
H1p	Cultural capital → Participation intention	0.051	0.070	0.724	0.469	Not accept
H1q	Cultural capital → Conservation intention	−0.052	0.033	−1.546	0.122	Not accept
H1r	Cultural capital → Promotion intention	0.150	0.070	2.127	*	Accept
H2a	Human capital → Landscape service cognition	0.016	0.040	0.395	0.693	Not accept
H2b	Natural capital → Landscape service cognition	0.024	0.029	0.821	0.412	Not accept
H2c	Physical capital → Landscape service cognition	0.084	0.041	2.044	*	Accept
H2d	Financial capital→ Landscape service cognition	0.181	0.040	4.511	***	Accept
H2e	Social capital → Landscape service cognition	0.118	0.039	3.039	**	Accept
H2f	Cultural capital → Landscape service cognition	0.208	0.047	4.435	***	Accept
H3a	Landscape service cognition → Participation intention	0.772	0.152	5.076	***	Accept
H3b	Landscape service cognition → Conservation intention	0.401	0.077	5.182	***	Accept
H3c	Landscape service cognition → Promotion intention	0.614	0.123	4.994	***	Accept

Note: ***, **, * indicate significance at 1%, 5% and 10% levels, respectively.

Second, concerning the impact of different types of livelihood capital on landscape service cognition, the paths H2c, H2d, H2e, and H2f are significantly established, and all have positive path coefficients, while the other paths are not significant. This suggests that physical capital, financial capital, social capital, and cultural capital exert a significant positive influence on landscape service cognition. Additionally, in terms of the effect of landscape service cognition on behavioral intention, the paths H3a–H3c are significantly established, with positive coefficients, indicating that cognition of landscape services has a significant positive effect on participation, conservation, and promotion intentions.

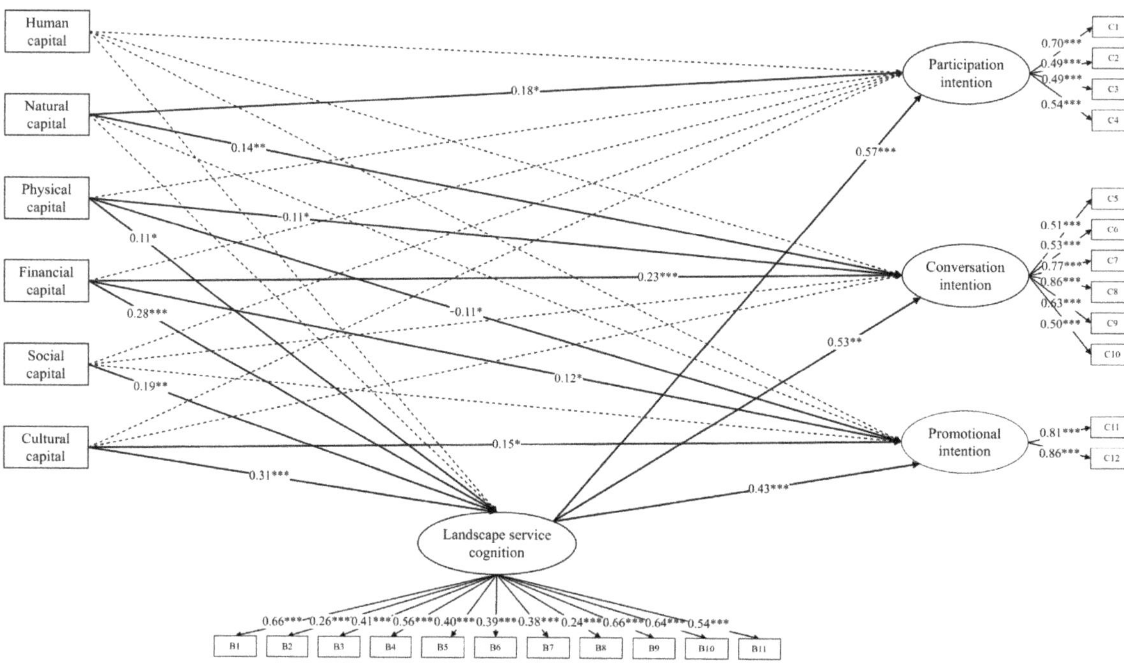

Figure 7. Results of the structural equation model regression. Note:. ***, **, * indicate significance at 1%, 5% and 10% levels, respectively.

In summary, farmers' natural capital, financial capital, and cultural capital stand out in this pathway, indicating their key roles in promoting farmers' active participation, conservation, and promotion behaviors.

This study conducted mediation effect analysis using the bootstrap test method, with 1000 repeated samples, and the results are shown in Table 7. The findings indicate that both physical capital and financial capital can enhance farmers' willingness to engage in conservation and promotion by increasing their cognition of landscape services. Cultural capital also has a positive indirect effect on promotion intention through the cognition of landscape services. However, the indirect effect of natural capital on participation and conservation intention through landscape service cognition did not reach significant levels, suggesting that other unrecognized pathways or factors may influence the effect of natural capital on farmers' behavioral intentions. Additionally, as shown in Tables 6 and 7, the direct and indirect effects of physical capital and financial capital exhibit consistency and significance across different behavioral intentions pathways, further highlighting the critical role these capitals play in shaping farmers' behavioral intentions.

Table 7. Mediation effect test results.

Hypothesis	Hypothesized Path	Standardized Indirect Effect	Bias-Corrected		Percentile		Result
			Upper Limit	Lower Limit	Upper Limit	Lower Limit	
H4	Natural capital → Landscape service cognition → Participation intention	0.026	0.088	−0.020	0.083	−0.038	Not accept
	Natural capital → Landscape service cognition → Conservation intention	0.024	0.082	−0.017	0.077	−0.021	Not accept
	Physical capital → Landscape service cognition → Conservation intention	0.060 *	0.129	0.011	0.123	0.007	Accept
	Physical capital → Landscape service cognition → Promotion intention	0.049 *	0.108	0.010	0.103	0.006	Accept
	Financial capital → Landscape service cognition → Conservation intention	0.146 **	0.226	0.073	0.232	0.077	Accept
	Financial capital → Landscape service cognition → Promotion intention	0.119 **	0.193	0.062	0.191	0.060	Accept
	Cultural capital → Landscape service cognition → Promotion intention	0.131 **	0.215	0.061	0.222	0.063	Accept

Note: **, * indicate significance at 5% and 10% levels, respectively.

5. Discussion

5.1. Differences in Landscape Service Cognition and Behavioral Intentions Among Farmers with Different Livelihood Strategies

Livelihood strategies refer to the action plans that farmers develop by leveraging and utilizing their resources to achieve livelihood goals or pursue livelihood outcomes [46]. To explore the differences in farmers' landscape service cognition and behavioral intentions, this study categorized respondents based on different livelihood strategies. Existing studies generally classify farmers' livelihoods using two main approaches: (1) the proportion of non-agricultural income to total income or the proportion of non-agricultural labor to total labor input [47], and (2) the sources of farmers' income or their employment status [48]. The first method is commonly used in economically developed areas or urban–rural integrated regions where comprehensive socioeconomic data are available, allowing for the accurate validation of farmers' income through data comparison. The second approach, based on income sources or employment status, is more applicable to regions with diverse industrial structures. For example, in this study, some villages have substantial industrial land, showcasing a development trend toward "residential areas integrated with handicraft industrial parks." Therefore, farmers' livelihoods rely not only on centuries-old irrigated agriculture but also on multiple income sources, including traditional handicrafts and processing industries. In this study, we classified the sample farmers into five livelihood strategies: diversified livelihood, wage-operated, purely agricultural, labor-led, and subsidy-dependent. This classification provides valuable insights for understanding livelihood diversity in regions with mixed industrial development.

The results of this study reveal significant differences in farmers' cognition of landscape services based on their livelihood strategies, particularly in the ecological and cultural dimensions. Zhang et al. found that these differences in ecological and cultural cognition are closely related to the type and degree of farmers' livelihood activities [36], which is further supported by our findings. Notably, in our study, diversified livelihood farmers demonstrated a much higher cognition of cultural value compared to those with other livelihood strategies. Most of these farmers hold public sector jobs, such as civil servants or village cadres, which gives them broader exposure to both agricultural and non-agricultural activities. This multidimensional participation enables them to better understand and appreciate the various values of landscape services, making them more reliant on traditional cultural resources and more committed to their protection and preservation [49]. In terms of the ecological dimension, while there were no substantial differences in the recognition of most ecological functions among the various livelihood strategies, wage-based farmers showed a significantly lower awareness of climate regulation compared to part-time equilibrium farmers. This finding aligns with studies by Wang et al. [50] and Guo et al. [51], which suggests that economically driven farmers tend to place less emphasis on ecological functions, focusing more on short-term economic gains rather than long-term ecological benefits.

Moreover, this study found that diversified livelihood farmers exhibited the highest levels of engagement in participation, protection, and promotion, particularly in ecological agriculture and tourism development. Farmers with diversified livelihoods are generally more willing to engage in regional activities and ecological conservation [24]. In contrast, subsidy-dependent farmers demonstrated a relatively low willingness to participate, which could be attributed to their heavy reliance on government subsidies and a lack of motivation for self-driven development [52]. In terms of protection intentions, diversified livelihood farmers also displayed a stronger willingness to engage in environmental conservation and the utilization of idle resources. This behavior may be driven by both their emotional attachment to rural heritage and their potential economic benefits. These farmers were also more inclined to contribute financially to the maintenance of rural heritage sites and the sustainable development of the environment. Regarding promotion intentions, diversified livelihood farmers showed higher enthusiasm than other groups, especially in educating and encouraging others, likely due to their greater social capital.

5.2. The Impact of Farmers' Livelihood Capital and Landscape Service Cognition on Behavioral Intentions

Farmers' livelihood capital and cognition of landscape services significantly influence their behavioral intentions. Livelihood capital, which includes human, natural, physical, financial, social, and cultural capital, collectively shapes farmers' decision-making when responding to different situations [53,54]. The results of this study indicate that different types of livelihood capital have varying degrees of impact on farmers' behavioral intentions and their understanding of landscape services.

First, natural capital was found to have a significant positive effect on farmers' willingness to participate in the protection of the landscape. This is consistent with the findings of Li et al., who demonstrated that the availability and quality of natural capital directly influence farmers' attitudes and behaviors toward cropland conservation measures [55]. When natural capital is abundant or in urgent need of protection, farmers are more inclined to participate in conservation efforts and seek appropriate compensation models. The positive effect of natural capital indicates that when farmers have access to sufficient natural resources, they are more inclined to participate in conservation activities to sustain both their livelihoods and the resources themselves. However, other studies suggest that in regions where natural resources are abundant and easily accessible, the impact of natural capital on farmers' behavior may diminish [56]. Additionally, cultural capital also had a notable positive impact on farmers' willingness to promote heritage conservation, further highlighting its importance in encouraging community participation and protecting heritage [13]. Similarly, financial capital had a clear positive impact on both the willingness to protect and promote, suggesting that greater economic capacity increases farmers' readiness to invest in rural heritage conservation and promotion. However, it is important to note that physical capital may have negative externalities. This study suggests that the accumulation of physical capital could actually decrease farmers' willingness to participate in protection and promotion efforts.

In contrast, this study found no significant effects of human capital and social capital on behavioral intentions, which contrasts with other research that highlights the importance of human capital in shaping environmental protection intentions [13]. This discrepancy could be due to the profit-driven nature of farmers' investment in human capital in the study area, where heritage conservation is viewed more as a public good. Furthermore, the influence of social capital varies considerably depending on its form—such as social trust or social networks—and the context in which it operates. In resource-limited environments, other types of capital, like natural and financial capital, tend to have a stronger influence on farmers' behavioral intentions, thereby diminishing the impact of social capital [57].

The study also found that landscape service cognition plays a key mediating role between livelihood capital and behavioral intentions. Physical, financial, and cultural capital indirectly increased farmers' intentions to protect and promote by enhancing their cognition of landscape services, which aligns with the findings of Yu et al. [58]. Although natural capital had some influence on landscape service cognition, its indirect effect on behavioral intentions through this cognition was not significant. Instead, it directly impacted farmers' participation and protection intentions. This result suggests that the influence pathways of natural capital are more complex and may involve additional mediating variables, which require further investigation.

Moreover, this study confirmed the direct and significant impact of landscape service cognition on farmers' behavioral intentions, particularly regarding their recognition of cultural and ecological functions, which significantly enhanced their willingness to engage in protection and promotion efforts. This aligns with the historical development of irrigation heritage landscapes, which have gradually evolved toward a more socio-ecological model.

5.3. Policy Recommendations

To address the previously discussed relationships between livelihood capital, landscape service cognition, and behavioral intentions, and to effectively enhance farmers' willingness to participate in, conserve, and promote rural heritage while achieving sustainable livelihood outcomes, this study recommends the following:

(1) Promoting the Synergistic Development of Cultural and Financial Capital.

This study highlights the crucial role that cultural capital plays in fostering farmers' willingness to engage in protective behaviors. Local governments can promote cultural tourism and festivals, while also creating cultural promotion platforms such as museums and traditional handicraft workshops. These efforts can increase farmers' cultural engagement and strengthen their sense of identity with local traditions. In doing so, not only can farmers' willingness to protect rural heritage sites be enhanced, but more diversified livelihood opportunities can also be created [41]. For diversified livelihood farmers, it is important to develop cultural resources and enhance the value-added aspects of cultural industries, allowing them to take a leading role in promoting sustainable livelihood outcomes. Financial capital, which has a significant impact on farmers' behavioral intentions, should also be addressed. Current policies tend to focus on singular subsidies, but financial support should be more diversified. For example, diversified livelihood farmers could benefit from targeted funding that enables them to balance diversified livelihood strategies with greater ecological responsibility [59]. For wage-operated and subsidy-dependent farmers, long-term ecological subsidies and development funds could be used to enhance financial capital, ensuring their livelihood security.

(2) Enhancing Livelihood Capital to Improve Landscape Service Cognition.

Enhancing farmers' physical and natural capital, while emphasizing the role of social networks within social capital, can significantly improve their understanding of landscape services and increase their behavioral intentions. The reasonable development of physical capital helps improve farmers' production conditions and quality of life. Policies should guide farmers in upgrading or establishing infrastructure (such as energy-efficient and environmentally friendly facilities, green buildings, and ecological agriculture equipment) to ensure the efficient use of resources and environmental sustainability [60]. Improving natural capital is also key to strengthening farmers' willingness to engage in conservation efforts. Providing training in ecological agriculture can help farmers use natural resources more efficiently, which not only increases their awareness of landscape services but also boosts their willingness to participate in ecotourism, biodiversity conservation, and other related activities. Expanding social networks plays an important role in diversifying farmers' livelihoods and enhancing their behavioral intentions. Through cooperatives and mutual aid organizations, farmers can benefit from knowledge sharing and resource integration, and this mutual support system can increase their participation and enthusiasm in heritage conservation efforts [61]. The expansion of social capital strengthens farmers' sense of social identity and cooperation, thereby increasing their willingness to engage in and promote rural heritage conservation [62].

(3) Differentiated Strategies for Farmers with Different Livelihood Strategies.

Diversified livelihood farmers, with their strong understanding of the ecological and cultural value of landscape services, along with their high willingness to participate, protect, and promote, are well-positioned to take on a leading role in rural heritage conservation. Therefore, the government and relevant organizations should further empower them by encouraging participation in rural tourism, cultural preservation, and ecological agriculture projects. Village cadres, such as party branch secretaries and village committee directors, play a critical role in this process. Given that village leaders in China are selected through party elections and direct village-level elections, their influence and continuity in office provide a stable foundation for long-term rural development initiatives. By actively involving village cadres—who are often re-elected based on their performance and leadership—these projects can be better managed and more effectively implemented, ensuring that local resources are efficiently utilized and that rural communities benefit from sustainable practices. Wage-operated farmers, who tend to have weaker cognition of ecological functions and focus more on short-term economic gains, should be guided towards rural heritage conservation through economic incentives such as subsidies. For example, establishing special funds to subsidize industries related to ecological protection, particularly ecotourism and green agriculture, could attract this group to participate in heritage protection and development. Additionally, enhancing their awareness of ecological values could gradually shift their focus toward long-term ecological benefits. For purely agricultural farmers, who show a strong willingness to engage in ecological agriculture, supporting their involvement in rural heritage protection through ecological agricultural projects would be beneficial. This could include providing agricultural training, technical support, and market promotion to improve their productivity while encouraging the adoption of sustainable farming practices that protect the environment. This would enhance their contribution to the protection of rural heritage sites. Labor-led farmers, whose livelihoods largely depend on off-farm work, are less involved in ecological protection and cultural activities. For this group, strengthening their social networks could increase their intention to participate, especially by encouraging them to engage in heritage conservation and cultural transmission during holidays or off-peak farming seasons, alongside the local community. Finally, subsidy-dependent farmers, who generally show low behavioral intentions and are less active in rural heritage protection, may benefit from improved social security measures to stabilize their livelihoods. At the same time, enhanced awareness campaigns could help increase their understanding of the cultural and ecological value of rural heritage sites.

5.4. Limitations and Future Works

This research uncovers important insights into the relationship between farmers' livelihood capital, landscape service cognition, and their behavioral intentions in rural heritage conservation. It offers both theoretical and practical guidance for improving sustainable livelihood strategies and heritage preservation practices. However, two limitations persist in this research. First, physical capital data remain difficult to obtain, requiring refined survey designs for more detailed information. Second, the influence of the industrial economy, natural environment, and cultural characteristics on farmers' behaviors has not been fully explored. Future research might delve deeper into these areas to better understand the feedback mechanisms involved.

6. Conclusions

The study of sustainable livelihoods for farmers in rural heritage sites holds significant importance for heritage conservation, transmission, and the sustainable development of farmers. This study constructs a research framework within the Sustainable Livelihoods Approach, integrating Cognition and Behavior Theory, and develops a scale to assess landscape service cognition and farmers' behavioral intentions. Using a sample of 371 farmers from representative villages within Putian's Ecological Green Heart, this study explores the differences among farmers with various livelihood strategies and analyzes key factors influencing their behavioral intentions from the perspectives of livelihood capital and landscape service cognition. The key findings are as follows:

(1) The normalized levels of human capital (0.541), social capital (0.671), and cultural capital (0.645) are relatively high among farmers in the study area, while the levels of natural capital, physical capital, and financial capital are comparatively low.
(2) Diversified livelihood farmers exhibit the highest levels of overall landscape service cognition and behavioral intentions. They stand out in their recognition of ecological and cultural services and demonstrate a strong enthusiasm for participating in ecological agriculture and the handicrafts and processing industries. In contrast, subsidy-dependent farmers have the lowest level of behavioral intentions.
(3) Natural capital, financial capital, and cultural capital play key roles in influencing farmers' landscape service cognition and behavioral intentions. Additionally, landscape service cognition mediates the relationship between livelihood capital and farmers' behavioral intentions.
(4) To enhance farmers' willingness to protect rural heritage, the focus should be on improving and accumulating natural, physical, and financial capital. In contrast, to increase farmers' willingness to promote rural heritage, the accumulation of cultural capital is particularly crucial.

Author Contributions: Conceptualization, S.L. and X.Z.; methodology, S.L.; software, S.L.; validation, S.L., Y.C. and J.C.; formal analysis, S.L.; investigation, S.L., Y.C. and J.C.; resources, X.Z.; data curation, S.L. and Y.C.; writing—original draft preparation, S.L.; writing—review and editing, S.L. and X.Z.; visualization, S.L.; supervision, X.Z.; project administration, X.Z.; funding acquisition, X.Z. All authors have read and agreed to the published version of the manuscript.

Funding: This research was funded by the National Natural Science Foundation of China, grant number 52308053; the Special Foundation of Fujian Provincial Department of Finance, grant number 202315; the Humanities and Social Sciences Research Youth Foundation, Ministry of Education, China, grant number 20YJC760079.

Data Availability Statement: Due to the privacy and confidentiality of the respondents, the questionnaire data used in this study cannot be made publicly available. However, the data can be accessed upon reasonable request and under the condition that participant privacy is ensured, by contacting the corresponding author.

Conflicts of Interest: The authors declare no conflicts of interest.

Appendix A

Figure A1. The master planning for the protection and utilization of the ecological green center.

References

1. Chambers, R.; Conway, G. *Sustainable Rural Livelihoods: Practical Concepts for the 21st Century*; Institute of Development Studies: Brighton, UK, 1992; IDS Discussion Paper No. 296.
2. Zhou, Z.; Gao, M.; Jiang, Y.; Li, H.; Shi, Y. Study on identification of rural settlement heritage aggregation areas in China and its spatial pattern. *J. Urban Reg. Plan.* **2023**, *15*, 1–24.
3. Young, J.C.; Calla, S.; Lécuyer, L. Just and sustainable transformed agricultural landscapes: An analysis based on local food actors' ideal visions of agriculture. *Agric. Ecosyst. Environ.* **2023**, *342*, 108236. [CrossRef]
4. Csurgó, B.; Smith, M.K. The value of cultural ecosystem services in a rural landscape context. *J. Rural Stud.* **2021**, *86*, 76–86. [CrossRef]
5. Li, H.; Xu, N.; Xie, X. On the efficiency and influencing factors of farmers' tourism livelihood in world heritage site: A case study of Danxia Mt. in Guangdong Province. *Sci. Technol. Manag. Land Resour.* **2021**, *37*, 41–51.
6. Su, M.M.; Wall, G.; Wang, Y. Integrating tea and tourism: A sustainable livelihoods approach. *J. Sustain. Tour.* **2019**, *27*, 1591–1608. [CrossRef]
7. Bires, Z.; Raj, S. Tourism as a pathway to livelihood diversification: Evidence from biosphere reserves, Ethiopia. *Tour. Manag.* **2020**, *81*, 104159. [CrossRef]
8. Yang, L.; Liu, M.; Min, Q.; Li, W. Specialization or diversification? The situation and transition of households' livelihood in agricultural heritage systems. *Int. J. Agric. Sustain.* **2018**, *16*, 455–471. [CrossRef]
9. Zhang, T.; Li, J.; Wang, Y. Effects of Livelihood Capital on the Farmers' Behavioral Intention of Rural Residential Land Development Right Transfer: Evidence from Wujin District, Changzhou City, China. *Land* **2023**, *12*, 1207. [CrossRef]
10. Bernstein, H.; Byres, T.J. From Peasant Studies to Agrarian Change. *J. Agrar. Change* **2001**, *1*, 1–56. [CrossRef]
11. He, S.; Li, H.; Min, Q. Value and conservation actors of Important Agricultural Heritage Systems (IAHS) from the perspective of rural households. *Resour. Sci.* **2020**, *42*, 870–880. [CrossRef]
12. Peng, Y.; Liu, W.; Xiong, C.; Bai, X. Value Transformation and Ecological Practice: The Path to Realizing the Value of Ecotourism Products in Heritage Sites—A Case Study of the Qitai Dry Farming System in Xinjiang. *Sustainability* **2024**, *16*, 5031. [CrossRef]
13. Zhang, C.; Min, Q.; Zhang, H.; Zhang, Y.; Tian, M.; Xiong, Y. Analysis on the rural households aiming at the conservation of agricultural heritage system. *China Popul. Resour. Environ.* **2017**, *27*, 169–176.
14. Scoones, I. *Sustainable Rural Livelihoods: A Framework for Analysis*; Institute of Development Studies: Brighton, UK, 1998; IDS Working Paper No. 72.
15. DFID. *Sustainable Livelihoods Guidance Sheets*; DFID: London, UK, 2000.
16. Simpson, M.C. An integrated approach to assess the impacts of tourism on community development and sustainable livelihoods. *Community Dev. J.* **2009**, *44*, 186–208. [CrossRef]
17. Wang, F.; Zheng, H.; Wang, X.; Peng, W.; Ma, D.; Li, C. Classification of the Relationship between Household Welfare and Ecosystem Reliance in the Miyun Reservoir Watershed, China. *Sustainability* **2017**, *9*, 2290. [CrossRef]
18. Yang, L.; Liu, M.; Min, Q.; He, S.; Jiao, W. Review of eco-environmental effect of farmers' livelihood strategy transformation. *Acta Ecol. Sin.* **2019**, *39*, 8172–8182.
19. Wu, J. Social Networks, Perceived Value and Farmers' Behavior of Cultivated Land Quality Protection. Ph.D. Thesis, Northwest Agriculture and Forestry University, Xianyang, China, 2021.
20. Chen, S. Mainstreaming Ecosystem Service Trade-Offs into Farmland Use and Livelihood Strategies. Ph.D. Thesis, Zhejiang University: Hangzhou, China, 2021.
21. Lei, J.; Liu, Y.; Gao, L. Research on the relationship between fishermen's willingness to exit fishing and sustainable livelihood capitals in the Yangtze River Basin: Based on survey data from the Poyang Lake Area. *Jiangxi Soc. Sci.* **2022**, *42*, 196–205.
22. Sun, F.; Ouyang, C.; Xu, X.; Jia, Y. Study on farmers' willingness to change livelihood strategies under the background of rural tourism. *China Popul. Resour. Environ.* **2020**, *30*, 153–160.
23. Zhang, Y.; Halder, P.; Zhang, X.; Qu, M. Analyzing the deviation between farmers' Land transfer intention and behavior in China's impoverished mountainous Area: A Logistic-ISM model approach. *Land Use Policy* **2020**, *94*, 104534. [CrossRef]
24. Liu, H.; Zhou, Y. Farmers' Cognition and Behavioral Response towards Cultivated Land Quality Protection in Northeast China. *Sustainability* **2018**, *10*, 1905. [CrossRef]
25. Jia, Y.; Zhao, M. The influence of environmental concern and institutional trust on farmers' willingness to participate in rural domestic waste treatment. *Resour. Sci.* **2019**, *41*, 1500–1512. [CrossRef]
26. Li, M.; Wang, J.; Zhao, P.; Chen, K.; Wu, L. Factors affecting the willingness of agricultural green production from the perspective of farmers' perceptions. *Sci. Total Environ.* **2020**, *738*, 140289. [CrossRef] [PubMed]
27. Li, W.; Qiao, D.; Hao, Q.; Ji, Y.; Chen, D.; Xu, T. Gap between knowledge and action: Understanding the consistency of farmers' ecological cognition and green production behavior in Hainan Province, China. *Environ. Dev. Sustain.* **2024**, 1–25. [CrossRef]
28. Wang, X.; Li, H.; Zhang, G. How does livelihood capital affect farmers' pro-environment behavior? Mediating effect based on value perception. *J. Agro-For. Econ. Manag.* **2021**, *20*, 610–620.
29. Lyu, X.; Liu, Y.; Peng, W. The association of farmers' cognition, intention and behaviour towards sustainable intensification of cultivated land use in Shandong Province, China. *Int. J. Agric. Sustain.* **2024**, *22*, 2318930. [CrossRef]
30. Lee, C.-H. Understanding rural landscape for better resident-led management: Residents' perceptions on rural landscape as everyday landscapes. *Land Use Policy* **2020**, *94*, 104565. [CrossRef]

31. Li, C.; He, G. The mechanism and empirical study of livelihood capital's impact on the differentiation of farmers in mountainous area under SLA framework—A case study of farmers in Guangxi mountainous area. *Chin. J. Agric. Resour. Reg. Plan.* **2021**, *42*, 144–156.
32. Cai, J.; Wu, X. The impacts of rural tourism on the vulnerability of farmers' livelihood: From the perspective of coupled social-ecological system. *Res. Agric. Mod.* **2018**, *39*, 654–664.
33. Wang, B.; He, S.; Min, Q.; Sun, Y. Influence of traditional ecological awareness of rural households on tourism livelihood options in agricultural heritage sites. *Tour. Trib.* **2024**, *39*, 98–111.
34. Shi, Z.; Zhang, H. Research on social, norms, environmental regulations and farmers' fertilization behavior selection. *Chin. J. Agric. Resour. Reg. Plan.* **2021**, *42*, 51–61.
35. Fleig, L.; Ngo, J.; Román, B.L.; Ntzani, E.E.; Satta, P.; Warner, L.M.; Schwarzer, R.; Brandi, M.L. Beyond single behaviour theory: Adding cross-behaviour cognitions to the health action process approach. *Br. J. Health Psychol.* **2015**, *20*, 824–841. [CrossRef]
36. Zhang, S.; Chen, L.; Hu, W.; Li, G. Mechanism of rural land landscape cultural value co-creation: Scenario, cognitions, and farmers' behavior. *Environ. Dev. Sustain.* **2023**. [CrossRef]
37. Hou, C.; Li, Z.; Xun, L.; Li, X.; Li, S.; Chen, X. Relationship between residents' place perception, place attachment and protection attitude: A case study of typical Dike-pond agricultural village in Foshan City. *Areal Res. Dev.* **2023**, *42*, 162–167.
38. Walker, A.J.; Ryan, R.L. Place attachment and landscape preservation in rural New England: A Maine case study. *Landsc. Urban Plan.* **2008**, *86*, 141–152. [CrossRef]
39. Wu, J.; Chen, S.; Zhou, K.; Chen, X. Influence of Livelihood Capital of Rural Reservoir Resettled Households on the Choice of Livelihood Strategies in China. *Water* **2022**, *14*, 4055. [CrossRef]
40. Zhang, X.; Lin, C.; Ren, W.; Wang, X. Study on the Landscape Multifunctionality and Multi-subject Game in the Fuzhou West Lake Water Heritage. *Chin. Landsc. Archit.* **2023**, *39*, 36–42.
41. Ma, X.; Wang, R.; Dai, M.; Ou, Y. The influence of culture on the sustainable livelihoods of households in rural tourism destinations. *J. Sustain. Tour.* **2020**, *29*, 1235–1252. [CrossRef]
42. He, Y. *Reclamation and Irrigation: Researching on History of Farmland Water Conservancy in Ancient Putian (627-1850)*; Nanjing Agricultural University: Nanjing, China, 2019.
43. Ren, L.; Zhang, M.; Chen, Y. The relationship between livelihood capital, multi-functional value perception of cultivated land and farmers' willingness to land transfer: A regional observations in the period of poverty alleviation and rural revitalization. *China Land Sci.* **2022**, *36*, 56–65.
44. Chen, J.; Pellegrini, P.; Wang, H. Comparative Residents' Satisfaction Evaluation for Socially Sustainable Regeneration—The Case of Two High-Density Communities in Suzhou. *Land* **2022**, *11*, 1483. [CrossRef]
45. Zhai, B.; Wang, Y.; Zhu, F.; Ji, T.; Xin, K. Differences in livelihood resilience of farm households in the Yellow River Basin under the background of livelihood strategies and its influencing factors: Taking Henan Province as an example. *Econ. Geogr.* **2024**, *44*, 156–165.
46. Xiang, H.; He, F.; Zhou, G.; Zou, M. Evaluation of the sustainable livelihood of farming households in traditional village tourism area: A case study of four typical traditional villages in Chenzhou City, Hunan Province. *Prog. Geogr.* **2023**, *42*, 884–897. [CrossRef]
47. Xiao, Y.; Yin, K. Impact of part-time farmer households' characteristics on farmer decision—Based on asurvey of 715 households in three gorges reservoir. *Resour. Dev. Mark.* **2015**, *31*, 1510–1514.
48. Zheng, D.; Huang, X.; Wang, C. Farmers' livelihood resilience and its optimization strategy in Loess Plateau of north Shaanxi province. *J. Arid Land Resour. Environ.* **2020**, *34*, 9–16.
49. Dou, Y.; Yu, X.; Bakker, M.; De Groot, R.; Carsjens, G.J.; Duan, H.; Huang, C. Analysis of the relationship between cross-cultural perceptions of landscapes and cultural ecosystem services in Genheyuan region, Northeast China. *Ecosyst. Serv.* **2020**, *43*, 101112. [CrossRef]
50. Wang, R.; Zhao, X.; Liu, J.; Wang, X.; Lan, H.; Xue, B. Dependence of farmers'livelihoods on environmental resource in key ecological function area: A case study of Gannan Plateau, China. *Chin. J. Appl. Ecol.* **2020**, *31*, 554–562.
51. Guo, A.; Wei, Y.; Zhong, F.; Wang, P. How do climate change perception and value cognition affect farmers' sustainable livelihood capacity? An analysis based on an improved DFID sustainable livelihood framework. *Sustain. Prod. Consum.* **2022**, *33*, 636–650. [CrossRef]
52. Ferse, S.C.A.; Liu, W.; Vogt, C.A.; Luo, J.; He, G.; Frank, K.A.; Liu, J. Drivers and Socioeconomic Impacts of Tourism Participation in Protected Areas. *PLoS ONE* **2012**, *7*, e35420.
53. Ren, J.; Lei, H.; Ren, H. Livelihood Capital, Ecological Cognition, and Farmers' Green Production Behavior. *Sustainability* **2022**, *14*, 16671. [CrossRef]
54. Tang, L.; Xu, Y.; Wang, W.; Wang, Y. Impact of livelihood capital and rural site conditions on livelihood resilience of farm households: Evidence from contiguous poverty–stricken areas in China. *Environ. Sci. Pollut. Res.* **2023**, *30*, 123808–123826. [CrossRef]
55. Li, G.; Fang, C.; Qiu, D.; Wang, L. Impact of farmer households' livelihood assets on their options of economic compensation patterns for cultivated land protection. *J. Geogr. Sci.* **2014**, *24*, 331–348. [CrossRef]
56. Yang, X.; Sang, Y.; Zhang, A. How livelihood capital affects farmers' willingness to pay for farmland non-market value? Evidence from Jianghan Plain, China. *Environ. Sci. Pollut. Res.* **2022**, *29*, 51456–51468. [CrossRef]

57. Liang, C.; Cheng, W. Study on the impact of livelihood capital on farmers' willingness of homestead exit—An empirical analysis based on ordered logit model in Dingxi City. *Nat. Resour. Econ. China* **2023**, *36*, 83–89.
58. Yu, C.; Xu, F.; Liu, Y.-M.; Pu, L.-J.; Liu, L.-Y.; Xu, Y. Analysis on farmers' behavioral intension of cropland use and its influencing factors in the coastal areas of Northern Jiangsu province. *J. Nat. Resour.* **2022**, *37*, 1643–1653.
59. Su, M.; Sun, Y.; Min, Q.; Jiao, W. A Community Livelihood Approach to Agricultural Heritage System Conservation and Tourism Development: Xuanhua Grape Garden Urban Agricultural Heritage Site, Hebei Province of China. *Sustainability* **2018**, *10*, 361. [CrossRef]
60. Li, T.; Tian, W.; Zhang, S.; Wang, S. Environmental regulation, high-quality economic development and ecological capital utilization. *Front. Environ. Sci.* **2024**, *12*, 1325289. [CrossRef]
61. Zhu, X.; Wang, G. Impact of Agricultural Cooperatives on Farmers' Collective Action: A Study Based on the Socio-Ecological System Framework. *Agriculture* **2024**, *14*, 96. [CrossRef]
62. Li, W.; Shuai, C.; Shuai, Y.; Cheng, X.; Liu, Y.; Huang, F. How Livelihood Assets Contribute to Sustainable Development of Smallholder Farmers. *J. Int. Dev.* **2020**, *32*, 408–429. [CrossRef]

land

MDPI

Review

How Public Urban Space Enhance Restoration Benefits Through Combined Multisensory Effects: A Systematic Review

Wen Dong [1], Donghui Dai [1], Pengyuan Shen [2], Rui Zhang [1] and Mei Liu [1,*]

1 School of Architecture, Harbin Institute of Technology (Shenzhen), Shenzhen 518055, China; dongwen@stu.hit.edu.cn (W.D.); daidonghui@hit.edu.cn (D.D.); e0703442@u.nus.edu (R.Z.)
2 Shenzhen International Graduate School, Tsinghua University, Shenzhen 518055, China; shenpengyuan@sz.tsinghua.edu.cn
* Correspondence: mei.liu@hit.edu.cn; Tel.: +86-13022059765

Abstract: Rapid urbanization has heightened attention fatigue and physiological stress among urban residents. However, high-density urban construction, mainly covered by streets, squares/plazas, and buildings, in particular, poses a challenge to traditional restoration dominated by green infrasturcture. This research aims to conduct a systematic review of the available body of knowledge regarding the relationship between public urban spaces and restorativeness based on combined multisensory effects. First, a conceptual framework was proposed to connect spatial properties of public realms, sensorial mechanisms, and restoration benefits. As a result, detailed spatial elements and organization affecting multiple sensory modalities combined via additive, antagonistic, and synergistic effects were extracted, which essentially act on urban restoration, including attention, stress, and emotional recovery. Last, but not least, compared with previous research, restoration design strategies of public urban spaces accelerating multisensory integration were drawn to contribute new insights for future high-density urban development.

Keywords: public urban space; multisensory perception; combined effect; restoration benefit; spatial property; urban design

1. Introduction

The city has become the main place of human habitation, and 68% of the population worldwide is expected to live in cities by 2050 [1,2]. Rapid urbanization has heightened attention fatigue and physiological stress among urban residents [3–7]. From the design perspective, restorative urban spaces provide appropriate spatial pathways to trigger the psychological and/or physiological recovery processes of resource-consuming individuals through particular urban settings, which thereby improves the whole well-beings but also the effective functioning of cities [8–11].

A traditional restorative environment normally indicates natural space dominated by green elements [12,13]. However, high-density urban construction captured by extensive hard surfaces and an abundance of built infrastructure poses a challenge to get close to nature or insert large-scale green spaces [14–18]. Under this circumstance, urban designers and researchers gradually paid their attention to the restoration benefits of public urban spaces with a more hardscape layout, such as streets and squares [16,19–22]. Several studies pointed out that well-designed and appealing public urban spaces might be as curative as natural environment [23,24]. For example, Lindal and Hartig found physical attributes of the urban residential street, especially the variation of building façades and building height, could significantly influence possibilities for people's restoration [25]. Zhao et al. detected that the clearness of painted traffic signs and diverse species of plants in the streetscape were capable of alleviating physical stress [22]. As for the research about squares/plazas, Subiza–Pérez et al. determined the composition and number of benches, sculptures, and playgrounds in urban squares are important in affecting restoration qualities, which functionally offer opportunities for people to rest and socialize [26]. Though research on the restoration of these urban environments is inadequate, it is still explicit that certain urban settings could arouse environmental perception formed by multiple sensations, including sight, touch, sound, etc., and the mediation of behavioral activities and emotional responses are beneficial to increase restorative effects accordingly [27].

Previous studies in terms of restoration benefits of urban spaces mainly focus on the account of their spatial properties and the visual stimuli. However, other sensory stimuli felt by the auditory, olfactory, gustatory, and tactile systems could also make key impressions on perceived restorativeness [28–30]. In fact, due to the rich mix of built elements and human activities, people in the urban environment are more easily exposed to multisensory stimuli rather than single senses [15,19,31]. Urban surroundings, such as air pollution odors, bird sounds, and various thermal factors, have been proven to positively or negatively affect restoration effectiveness [32–35]. Nevertheless, minimal research has tried to reveal the complex mechanism between multisensory effects on urban settings and the corresponding restoration benefits. Ha and Kim investigated the combination of visual and auditory influence within a biodiverse environment, which could provide restorative effects for students' everyday life on campus [36]. Also, Jiang et al. conducted a laboratory experiment to analyse how odour affects people's satisfaction with urban landscapes by modulating their noise and visual perception simultaneously [37]. Though these studies partly verified the Gestalt principle that different sensory inputs affect environmental perception as a whole, recovery outcomes, depending on the mode, approach, and level of multisensory stimulation, are still lacking [38,39].

In summary, despite natural spaces with large green substrates, public spaces primarily constituted by hard materials, by means of their abundant physical traits and two- and three-dimensional layouts, also could conduct multiple sensory experiences and accelerate human health. Therefore, this study aimed to provide a systematic overview summarizing, analyzing, and understanding the relationship between restoration benefits and the spatial properties of public urban spaces (e.g., street and square) via combined multisensory effects. To achieve this objective, the following two questions need to be answered: (1) Which spatial character and spatial organization of public urban spaces could effectively act on restoration benefits? (2) How do these spatial properties influence recovery through different multisensory combination patterns and effects?

This paper was divided into the following sections. In Section 1, a theoretical framework was constructed to illustrate the linkage among the spatial property of public urban spaces, multisensory effects, and restoration benefits. Section 2 describes the methodology of this systematic review, which specifies the inclusion and exclusion criteria of data collection and how the body of evidence was assessed. In Section 3, the reviewed studies were grouped by which the inner relationship between various multisensory combination patterns (i.e., visual-auditory, olfactory-auditory, tactile-visual, olfactory-thermal, tacitle-olfactory, visual-olfactory, thermal-visual, and tri-sensory combinations) and main categories of restoration benefits, including attention, stress, and emotional recovery, were summarised, respectively. In Section 4, comparisons were made with previous research on the restorative effectiveness of natural spaces and single senses' efficiency, as well as a series of design strategies for future urban restoration practices.

2. A Framework Linking the Spatial Property of Public Urban Spaces, Multisensory Effects, and Restoration Benefits

Public urban spaces, including streets, squares/plazas, building exterior spaces, and roof spaces, etc., play an important role for people's daily use [40]. In addition to integrated green spaces, these everyday urban environments provide the residents' unnoticed existing situation and frequent embodied experience [41]. Based on Affordance Theory [42] and Environmental Perception Theory [43], an individual's cognition is significantly influenced by spatial characteristics. Environmental factors offer specific "affordances" through various sensory inputs, which are processed through sensation and perception to influence individuals' restorative experiences [44,45]. A large number of studies of restorative urban environments have stated that contact with certain spatial features may effectively facilitate people's attention, stress, and emotional recovery, while this psychological response is immediate, spontaneous, and accompanied by a series of positive emotions [31,43,46–50]. Therefore, in order to grasp the nature of how public urban spaces influence people's restoration benefits, it is useful to start extracting spatial clues in terms of the composition and organization of urban elements and multisensory experiences incorporated [51,52]. We conducted a framework linking spatial properties of public urban spaces, multisensory effects, and restoration benefits in Figure 1.

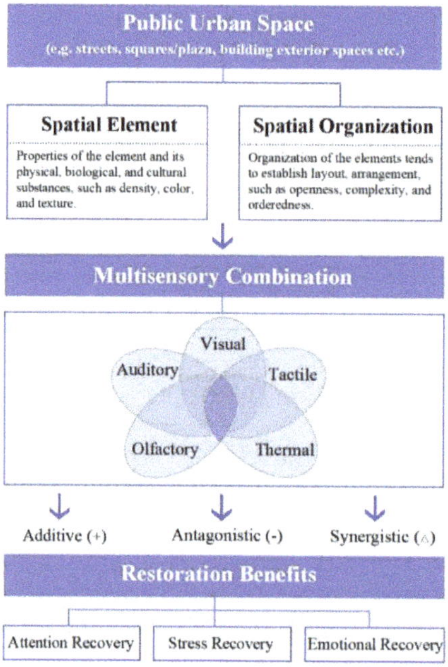

Figure 1. A framework linking spatial properties of public urban spaces, multisensory effects, and restoration benefits.

Firstly, properties of the element and its physical, biological, and cultural substances, such as density, color, and texture, could indicate straightforward recovery effects. For example, the façade design of buildings based on the selection of materials or vertical greening, and the underlying/in-house commercial facilities emanating artificial aromas of fragrance, coffee, bread, etc., could lead to different aesthetic and tactile perceptions, which provide calming and recovery effects for people who work nearby or just pass through the neighbourhood [25,53,54]. Moreover, the density of street trees has been observed to be an essential design and planning factor in the streetscape in terms of attention and stress restoration through creating green vista, while the compactness of fragrant plants that affect the odor concentration has been shown to reduce human's stress [22,55,56]. Secondly, the organization of the elements tends to establish layout, arrangement, and spatial relationships among multiple urban elements to indicate perceptual experiences, such as openness, complexity, and orderedness [57–59]. Apart from visual aesthetics, it could also induce other sensory stimuli ultimately acting on recovery benefits [59–62]. For example, people prefer to experience a feeling of openness in high-density commercial and business areas, which provoke positive responses by allowing odor dispersal and reducing noise levels [49,63,64]. Furthermore, the complexity of planting design could effectively create a microclimate that promotes better thermal satisfaction and the attenuation of artificial odors [63,65,66]. Also, the street ratio (i.e., H/W) and tree plantings with different canopy sizes could functionally determine thermal and visual comfort and ameliorate heat stress for urban dwellers [67].

As mentioned, the spatial properties, including variations of texture, form, pattern, and colour of urban elements, initiate people to synthetically perceive the environment. However, different sensory inputs stemming from the outdoor space cannot simply be regarded as the sum of visual, olfactory, auditory, tactile, and thermal senses. All senses collaborate to grasp information through complex interactions with the surroundings and influence various emotional responses [39,68]. Based on the previous research, three key multisensory combinations acting on restorativeness can be summarized, namely additive, antagonistic, and synergistic effects [69–71].

- The additive effect refers to several sensory factors working together to promote recovery by providing a favorable perceptual experience. For example, rich plant communities with fragrant plant odours could promote restoration benefits [72]. The combination of green landscapes and birdsong has also been shown to promote a favourable perception [54].
- The synergistic effect is manifested by the fact that one sensory sensation can compensate for a certain sensory discomfort, which in turn promotes recovery. Olfactory and visual experiences in urban spaces have been shown to increase the overall perception by alleviating thermal discomfort from high temperatures, as well as easing the annoyance of traffic noise [33,37]. Comfortable sensations of olfactory and tactile could similarly promote recovery in an environment with low green visibility [72,73].
- The antagonism effect is the inhibitory effect of sensory factors on recovery, and the poor performance of one sensory factor may exert a masking effect on others. A masking effect has been shown to exist between sound and smell, where the introduction of traffic sound may reduce overall perception when the sense of smell is comfortable [74]. The tactile experience of crowdedness may also reduce the effectiveness of other sensory experiences [75].

3. Materials and Methods

3.1. Data Collection

The data collected come from the Web of Science (WOS) core collection database. WOS is the premier research platform for information in several scientific fields and the world's most reliable publisher-independent global citation database. To gain useful clues, we divided the topic into three main categories as obligatory terms ("urban public space", "multisensory", and "restoration benefit"), with each of them having a set of keywords for their related concepts. Meanwhile, we excluded search terms such as forests, villages, and suburbs, which do not meet the definition of public urban spaces in this study.

- Public urban space: TS1 = (public urban space* OR urban OR high-density city OR square OR street)
- Multisensory: TS2 = (multisensory OR multi-sensory OR multisensory OR audi* OR sound* OR acoustic OR noise OR smell OR olfactory OR fragrance* OR tactile OR touch OR haptics OR thermal)
- Restoration benefits: TS3 = (restoration benefit* OR restoration environment* OR therapeutic landscapes OR stress relief OR stress recovery OR attention recovery OR mood states OR emotional recovery)
- Non-urban space: TS4 = (forest* OR village OR outskirts OR suburb OR wilderness)

We then matched these three groups of keywords, with each as (TS1) AND (TS2) AND (TS3) NOT TS4. Data were accessed on April 2024 from the Harbin Institute of Technology Online Library.

3.2. Literature Screening

The results yielded 447 documents from the WOS. The PRISMA diagram illustrates the search process [12] (Figure 2). Literature from unrelated disciplines was removed, such as dentistry medicine and materials science, and then 119 documents were obtained for a second round of screening (n = 119). Single sensory studies and literature only related to the natural environment were excluded, as well as literature not related to spatial elements and organisation (n = 70). The titles and abstracts of the studies met the inclusion criteria and were reviewed in full text (n = 49). Then, a further 19 papers were excluded because the literature focused on environments with large green substrates or built indoor environments were not accessible. In particular, the documents with term "park" were scrutinized. Specifically, references to squares, roadside spaces, or community parks as distinct from large-scale green spaces were included in the examination. Finally, 30 articles remained as the final set of references used in this study. The table resume of the studies selected for the full-text review is displayed in Table 1.

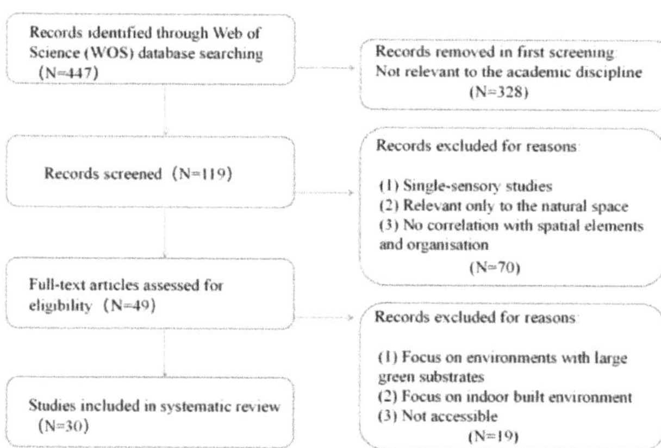

Figure 2. PRISMA flow diagram for paper screening.

Table 1. Literature review summary table.

Author, Time	Multisensory Combination	Measurement Method(s)	Study Site(s)
Jeon et al. (2021) [76]	Visual–Auditory	Narrative interviews; Questionnaire	Seoul, South Korea
Liu et al. (2022) [77]	Visual–Auditory	Short-version revised restoration scale (SRRS)	Harbin, China
Gallou et al. (2022) [78]	Visual–Olfactory–Tactile	Perceived restoration scale	England
Ha and Kim (2021) [36]	Visual–Auditory	Restorative state scale (RSS); Profile of mood states (POMS)	America
Ruotolo et al. (2024) [79]	Visual–Auditory	Emotional appraisal questionnaire; Cross-modal attention task; MAIA questionnaire	Italy
Jeon et al. (2023) [49]	Visual–Auditory	Heart rate variability (HRV) and electroencephalogram (EEG); Positive and negative affect schedule (PANAS)	South Korea
Ba and Kang (2019) [55]	Olfactory–Auditory	Questionnaire	Harbin, China
Chung et al. (2022) [80]	Visual–Auditory	Questionnaire	Hong Kong, China
Chang et al. (2023) [81]	Olfactory–Thermal	EEG measurements; Questionnaire	Xi'an, China
Pitt (2014) [82]	Tactile–Visual	Semi-structured interviews	Wales, UK
Ba et al. (2022) [83]	Olfactory–Auditory	Questionnaire	China
Schnell et al. (2016) [84]	Olfactory–Auditory–Tactile	HRV; Questionnaire	Tel Aviv, Israel
Lee et al. (2024) [73]	Tactile–Visual	Semi-structured interviews	Australia

Table 1. *Cont.*

Author, Time	Multisensory Combination	Measurement Method(s)	Study Site(s)
Jiang et al. (2021) [31]	Visual–Auditory	Multi-dimensional mood questionnaire (MDMQ)	Illinois, UK and Hong Kong, China
Lyu et al. (2022) [85]	Visual–Auditory–Thermal	Cognitive performance tests; Skin conductance level (SCL); Questionnaire	Sydney, Australia
Lyu et al. (2023) [50]	Thermal–Visual	Cognitive performance tests; Questionnaire	Sydney, Australia
Masullo et al. (2021) [86]	Visual–Auditory	Questionnaire	Hong Kong, China and Aversa, Italy
Wang and Zhao (2020) [87]	Thermal–Visual	Questionnaire	Nanjing, China
Elsadek et al. (2019) [33]	Thermal–Visual	Questionnaire	Shanghai, China
Ho and Au (2021) [88]	Visual–Auditory	Questionnaire	Hong Kong, China
Zhou et al. (2023) [89]	Visual–Auditory	Semi-structured interviews	Beijing and Huainan, China
Henshaw (2011) [90]	Olfactory–Visual	Semi-structured interviews	Britain
Zhang et al. (2023) [75]	Tactile–Visual	Galvanic skin response (GSR)	Salzburg, Austria
Zhou et al. (2023) [72]	Olfactory–Visual	HR; EEG; Cognitive performance tests; Questionnaire	Hangzhou, China
Cui et al. (2023) [91]	Olfactory–Auditory	EEG; HR; Blood pressure(BP); Questionnaire	Hangzhou, China
Osborne (2022) [34]	Tactile–Olfactory	Video elicitation interviews; Biosensing	Birmingham, UK
Xu et al. (2024) [19]	Visual–Auditory	Questionnaire	Hong Kong, China
Zhang et al. (2021) [92]	Thermal–Visual	Questionnaire	Xi'an, China
Karmanov and Hamel (2008) [23]	Visual–Auditory	Profile of mood states (POMS)	Sporenburg, Netherlands
Chan et al. (2021) [53]	Visual–Auditory	ECG; Questionnaire	Singapore

4. Results

The number of publications reviewed in this study is shown in Figure 3. The first relevant literature was published in 2008, surpassing the original nature versus urban dichotomy, which conducted studies in terms of visual-auditory perception of human health issues. The increasing trend of articles during the first decade was slow (four articles). However, the number of papers grew rapidly after 2018, with 87% of the total studies published in the last 5 years. The patterns and distribution of multisensory combinations are displayed in Figure 3. Studies with two and three types of combinations were included, with the former having the highest number. Among these studies, visual–auditory studies accounted for the largest share, followed by visual–thermal, olfactory–auditory, and tactile–visual studies. In addition, sight and smell are sensory factors that combine most frequently with other senses.

In the screened literature, minimal research was conducted in China (14 publications, 47%), followed by Britain, Australia (3 publications, 10%), and Italy, South Korea, and the Netherlands (2 publications, 7%), with some studies from the United States, Singapore, and Denmark (1 publication, 3%). It can be seen that studies are mainly concentrated in countries with a high degree of urbanization and high-density cities. Additionally, this research was related to multi-disciplinary subjects, especially including landscape design, public health, and environmental psychology. In the following sections, a comprehensive overview of how spatial elements and organization in public urban spaces influence restorative benefits was provided based on the classification of sensory types and the number of sensory combinations.

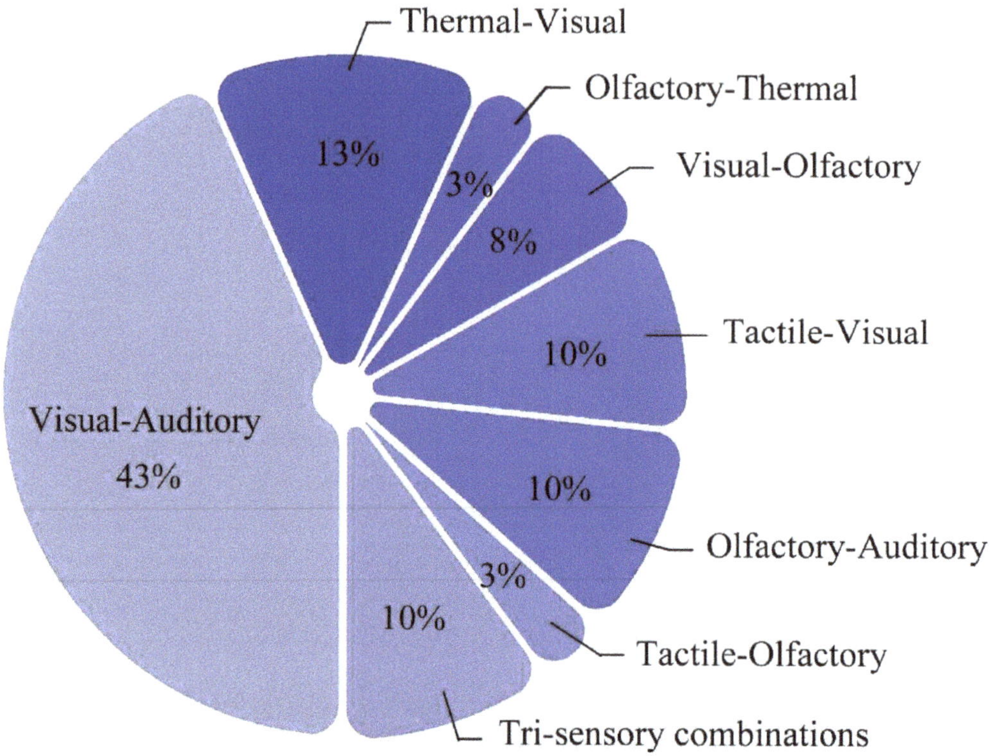

Figure 3. Patterns and distribution of multisensory combinations.

4.1. Auditory–Visual Perception

Studies in terms of auditory–visual perception accounted for the largest proportion of the literature screened (n = 13). Compared with previous research on the single auditory sense, which mainly pointed to the positive effect of sound types on recovery (e.g., bird calls and the sound of running water) and negative effects on mental health, (e.g., traffic noises might cause increased stress levels), auditory–visual perception research examined how to visually compensate for an existing urban background sound to promote restoration benefits [93–95]. More attention is paid to the coordination of different senses and the impact of everyday life scenarios.

Chung et al. conducted audio–visual stimuli via immersive virtual reality (IVR) combined with physiological devices, by which they found well-maintained building facades and how the coverage of vertical greening could provide urbanites stress relief under the exposure of noise levels from 55 to 65 dBA [80]. Meanwhile, Chan et al. also emphasized the role of vertical greening in buffering stress and promoting recovery under the background of a high-density noisy downtown area through a virtual cityscape environment facilitating the user's visual senses and body movement [53]. Ruotolo et al. revealed the use of colored architectural elements (e.g., colourful bridges) in open spaces had a positive impact on the well-being of the elderly, even in the presence of traffic sounds [79]. In addition to the vertical architectonic attributes, Chung et al. explored horizontal spatial organization, such as separation distance [80]. The presence of building spacing in the neighbourhood would significantly affect human perception of oppressiveness and decrease noise annoyance as well. A VR study combining narrative interviews with questionnaires found that open views strengthen the psychological effect of reducing perceived loudness [76].

Much research has identified that natural elements in urban surroundings notably help to achieve audio–visual coherence. The importance of vegetation for providing a green environment and its natural sounds (e.g., bird song, rustle of leaves) is well known [19,31,36,79]. Furthermore, as one of the most basic spatial elements, water stimulates different sensory effects and plays an important role in encouraging restoration benefits. For example, Masullo et al. pointed out that installations with water sound could effectively help people achieve attentional recovery from road traffic noises, while from the visual perspective, the enveloping shape and transparency of the water elements also influenced restoration [86]. Also, Jeon et al. have proved the proper design of audio-visual design elements could improve restoration response [49]. They realized that physical parameters like grey ratio (grey/urban element in a scene) were crucial to reinforcing acoustically calm and tranquil soundscapes. However, visual and auditory senses need to be properly coordinated to achieve satisfied perception. One study found that reducing plant diversity enhances the restoration of hydroacoustic sound in highly urbanised coastal areas, which may relate to the fact that landscape complexity reduces other perceptual stimuli [77].

Apart from the above-mentioned natural sounds, soundscapes derived from human daily life also exert an impact on recovery. Ho and Au marked the facilitating effect of street music performances on people's attention recovery, especially while walking along the street or pausing in the square [88]. Then, the behavioral responses of the audience and crowd size were also critical in enhancing the atmosphere [96]. Another study found that daily exercise routines with music like Tai Chi and dances for older people had an effect on the restoration of various groups, either ones who participated or stood by [89].

4.2. *Olfactory–Visual, –Auditory and –Thermal Perception*

Mutisensory studies related to the sense of smell accounted for 23% of the total screened literature, which mainly explains how restoration benefits are facilitated by exploiting compensatory or additive effects between olfaction and other senses.

Firstly, as for the olfactory–visual approach, Zhou et al. measured the effects of smelling materials, including gardenia, cedarwood, and no-fragrance vegetation on attention, stress, and emotional recovery processes at different levels of visible green [72]. The results showed that, in under 10–35% of greenery, only the fragrance of gardenia could significantly reduce human stress, while both fragrances of gardenia and cedar leaf produced stress-reduction effects in under 35–60% of visible greenery. In addition, cedar was more effective in attention recovery, and gardenias improved negative moods to a large extent. Henshaw pointed out that even small-scale plant landscapes could trigger a restorative effect after experiencing poorly perceived odors through positive alliesthesia [90]. The additive effect of sight and smell on recovery studies was also mentioned. For instance, rich plant communities combined with aromas could create a good restorative environment [72].

Secondly, interactive sensory stimulations between olfactory and thermals also have been proven to influence the perception of restoration. Chang et al. revealed that increasing scent comfort can alleviate the discomfort from summer heat, which results in stress relief and improved mood states [81]. They examined fragrance stimulation in different temperature stages, by which odour affected attentional recovery within an environment of 30.80 °C ≤ PET (physiological equivalent temperature), <44.53 °C rather than 44.53 °C ≤ PET < 58.27 °C, while emotional recovery is always present under different PETs.

Third, appropriate odors in public spaces can improve people's assessment of background noise. Cui et al. used physiological equipment and the profile of mood-state (POMS) scales to find that the odor stimulus of gardenia had a stress and emotional restorative effect under the sounds of everyday life and machinery at the sound pressure level of 55 db [91]. Ba and Kang conducted a questionnaire survey on-site to detect that the natural aroma of plants can reduce noise annoyance, while a double row of fragrant trees has a stronger effect on restoration benefits than single-row streets from the visual perspective [55]. Furthermore, food odors on both sides of the pedestrianized streets were shown to have the same effects [83].

4.3. Thermal–Visual Perception

Previous research about microclimate or thermal comfort of urban environments has pointed out that air temperature, heat-stress index, and wind speed are linked with physiological stress in humans. Searching multisensory approaches related to thermal perception, most of the existing studies focused on the thermal–visual synergistic effects, which explained how spatial characters and organizations modify the experience of visual and thermal environments and determine recovery. For example, Lyn et al. conducted a multisensory virtual reality to simulate dynamic environmental conditions, which noted that different shading patterns through different structures of space could facilitate restoration by providing landscape characters and optional thermal adaptation opportunities [50]. Another field study along urban roads by Elsadek et al. found that cherry trees had a relatively large sky view factor (SVF), but their aesthetic and symbolic significance still provided an equally powerful recovery experience like highly shaded roadside trees [33]. However, the inconsistency of the thermal–visual environment through seasons might influence the level of stress recovery [92]. Wang and Zhao found that the effect of deciduous trees on perceived recovery differed in spring and winter because of the visual penetration and sunlight exposure for the changes in permeability [87]. However, the inconsistency of the visual–thermal environment may reduce stress recovery when exposed to cold environments in images with high green visibility [92], though the presence of verdant vegetation may increase thermal comfort and overall perceptual effects by enhancing visual comfort [97,98].

4.4. Tactile–Visual and –Olfactory Perception

Multisensory studies related to the sense of tactile accounted for 13% of the total literature. Compared with the single-sense procedures focusing on the therapeutic function of different materials, such as sand, leaves, and water [99], the existing multisensory research has paid more attention to the mechanism between tactile and other sensory experiences, which were simultaneously stimulated by spatial properties of urban environments. Lee et al. explored dynamic visual features of green roofs (e.g., traditional lawn, biodiverse meadow, colourful succulent, and structural roof) and encouraged virtual visits to facilitate appropriate multisensory experiences in cities [73]. As for the results, they found that the significance of haptic experience arises when the space lacks complex and interesting landscape features. In addition, tactile also works with other sensations through additive or antagonistic effects on recovery. For example, Pitt demonstrated that, in community gardens, skin contact with a variety of plant textures and the whole-body tactility of horticultural activities like weeding and digging could contribute to an immersive experience with attention and stress recovery [82]. Zhang et al. conducted an outdoor experiment with wearable sensors and noted that the installation of roadside sitting facilities in a busy pedestrianized area or a narrow street might lead to increased levels of physiological stress owing to the close contact with people [75]. In addition to direct physical contact, our bodies also react to other atmospheric measures, such as air pollution. Osborne suggested that some microspatial features, such as wider pavement sidewalks, could avoid close proximity to polluted and turbulent air, while strategic vegetation was also useful to buffer pollutant flows [34].

4.5. Tri-Sensory Combinations

One study pointed to the role of the multisensory qualities (e.g., smells, sights, sounds, or textures) of historic buildings in restoration [78]. Their fractal patterns from building structures to decorative elements could add visual coherence and textured feeling, and the unique odor of urban history can trigger collective memories, which thus provoke recollections, emotional attachments, and associations [99–101]. The symbolic narrative of multisensory environments creates a sense of "moving to a bigger world" and leads urban dwellers to a feeling of being away from their stressful lives. What is more, two studies investigated the weight of the sensory contribution to restorative effects. To be specific, a representative study conducted in Singapore with individual sensory pleasantness votes and physiological measurements indicated thermal and visual pleasures were correlated to physiological recovery. However, the effect of auditory was not that significant [85]. In contrast, Schnell et al. did an outdoor experiment in the city center of Tel Aviv, Israel, and used portable measuring equipment to calculate in situ climatic and environmental variables [84]. They found that noise and crowding from the presence of people are more of a stressor than thermal load and carbonic oxide. It can be seen that different seasonal conditions and sensory combinations lead to different conclusions.

5. Discussion

5.1. Multisensory Effects of Restoration Benefits Between Public Urban Spaces and Natural Green Spaces

As for the restorative spatial property, the proportion of green elements is always positively correlated with restoration effects [77]. However, in high-density urban spaces, the presence of greenery is more effective for recovery than its area and density from time to time. To be specific, a modest amount of greenery is enough to ensure restoration effects through the critical role of auditory, olfactory, and/or thermal sensations [33,53,90]. Moreover, some architectural elements also show significant correlations with restorativeness in urban spaces than in naturalized environments. For instance, fountains have been proven to have a more pronounced restoration effect within urban surroundings enclosed by buildings [77]. Moreover, compared with the preference for high shelter spaces in natural restorative environments, people tend to demand more openness in compact public urban spaces. This might be because the open sky provides "soft glamour" [61], and open views have restoration benefits by reducing the perception of loudness and crowding, which could effectively facilitate urban background stressors [49,64,75]. Last but not least, "serene" is an important restorative quality of natural space [102,103]. However, everyday activities conducted in high-density urban spaces, such as streetscapes or squares, based on amenities people can sit in and stay in, and acoustic comfort from human-related sounds in plazas, can shape familiar and reassuring social atmospheres through multiple senses [88,89,104,105].

Regarding the multisensory mechanisms, relevant research in nature usually focuses on the additive effects in terms of spatial properties, since those restorative sources are mostly positive healing elements like water and vegetation [36]. Meanwhile, contexts in cities with more artificial and diverse urban elements, as well as interactions among multisenses, are relatively antagonistic and synergistic for recovery.

5.2. Comparison with the Previous Single-Sensory Studies

Compared with existing studies focused on the effects of individual senses on restorativeness, multisensory research provides a more comprehensive understanding of the complex mechanisms of sensory experience in a more realistic and applicable way.

(1) Compared to single-sensory studies, multisensory research can further clarify the effective stages of restoration. Previous single-sensory studies have often reported positive correlations between restoration benefits and factors such as plant cover and aroma concentration [77,106]. However, our study suggests that when a certain sensory input reaches a threshold, it may overshadow the positive effects of other senses [74,75]. For example, while single-sensory studies may highlight the benefits of increased plant coverage for psychological restoration, they overlook how excessive olfactory or auditory stimuli can diminish this effect when combined.

(2) Adjusting only one sensory input is not always sufficient to enhance recovery outcomes, as a significant restorative effect may emerge only through multisensory approaches. Furthermore, the limitations of single-sensory studies become more

evident in contexts with conflicting sensory inputs. Prior studies have shown that the link between urban biodiversity and restoration is often weak [107,108]. However, the addition of natural sounds has been found to significantly enhance the restorative experience [36]. In addition, while previous single-sensory studies emphasize the value of benches in promoting restorativeness by providing opportunities to pause [109,110], seating areas located near busy roads or crowded commercial zones are observed to lead to perceptions of overcrowding and increased stress in the multisensory research [75].

(3) Focusing on sensory coherence rather than single-sensory inputs is essential for promoting restoration [19,36,92]. While previous single-sensory studies have highlighted the significant role of plant aromas in recovery [56,111], they may become ineffective if they clash with the surrounding environment, even in areas with high green visibility [72].

5.3. Application of Combined Multisensory Experience on Restoration in Urban Design

Implementing restorative design strategy in urban practices involves integrating principles and techniques that promote the physical and mental well-being of city dwellers. In the past, this approach normally emphasizes utilizing certain sensory factors to mask the same sensory discomfort in urban environments [112–114]. However, it has been discovered that sounds like water and birdsong in noisy traffic environments are limited for noise masking and stress recovery [115,116]. Our research suggested that since recovery can be achieved through the synergic effect of multiple senses, restorative design needs to consider how spatial elements and their properties can effectively stimulate multisensory experiences and make good use of their interactions. By integrating sensory elements such as sight, sound, touch, and smell, designers can create more attractive and comfortable environments. This not only helps to enhance the psychological and physiological health of users but also promotes deeper emotional connections and community interactions.

Specifically based on the review, through sophisticated spatial design, odors could productively help to promote recovery in high-temperature, noisy, or low-green-visibility conditions in urban spaces [55,72,81]. Combining fragrant plants or the aroma of food with rich visual elements can enhance olfactory–visual perception and accelerate restoration. For example, London's Southbank Centre creates an immersive restorative experience by combining the aroma of coffee and bread from food stalls with visual elements such as murals and sculptures. In addition, combining essential oils from aromatic plants with landscape spray is an effective approach to integrating auditory, visual, and tactile experiences. Last but not least, using spatial organization thoughts, urban wind corridors based on the consideration of building spacing and spatial configuration can also help evacuate car exhaust fumes, reduce noise, and shape a comfortable thermal environment at the same time.

In highly urbanized areas, water installations can create dynamic, flowing landscapes that enrich multisensory experiences, thereby alleviating the annoyance and stress caused by urban noise [72,80,86]. For example, the forecourt of Sheffield Railway Station in the UK uses various water features to mask traffic noise, complemented by stainless-steel sculptures to create a restorative space. What is more, complex textures could be used to create different flow states, such as flips, drops, and bubbles, which provide rich visual, auditory, and tactile experiences.

It is also worth mentioning that urban outdoor space is a non-homeostatic environment, where an appropriate level of stimulation from other senses, even if it is a small change, is considered to be immediate and effective after discomfort is experienced [117]. This compensatory approach is based on the theory of "alliesthesia" [85,117,118]. Practically, when people pass through areas with car exhaust fumes or high solar radiation, small adjustments can help enhance restorative benefits, such as adding adjustable shading structures, using ground paving with varied textures, or placing small clusters of aromatic plants.

Future research and practice could further explore how to effectively integrate multisensory restorative design strategies into urban policies and planning regulations, such as "Green Building Standards" and "Healthy City Planning". These frameworks provide evaluation standards for health design. However, they often lack comprehensive consideration of multisensory experiences, addressing only limited sensory factors such as sound and air quality in certain indicators. Specifically, public spaces could gain additional points in green certification by incorporating multisensory elements, such as natural sounds, aromatic plants, or diverse tactile materials. Additionally, urban management should consider broader sensory needs when formulating policies under Healthy City initiatives. For example, historic buildings should be paid attention to and maintained to preserve the distinctive odor of cities that triggers collective memories. The cultural information about historic buildings could also be conveyed through added voice-over narration, which has been shown to achieve better restorative effects by combining visual and auditory elements [23].

6. Conclusions

Due to the limited accessibility to nature and over-exposure to urban environments, spatial design towards the psychological and physiological recovery of urban inhabitants becomes problematic. This research provides a comprehensive overview that tends to summarize, analyse, and understand the relationship among the spatial property of public urban spaces, multisensory effects, and restoration benefits from the perspective of urban planning and design. First, a conceptual framework was conducted to assist in organizing the available evidence, specifying key categories, and identifying the linkage between urban settings and restorativeness. Then, based on different sensory combinations via additive, antagonistic, and synergistic effects, detailed spatial elements and organizations acting on urban restoration were extracted. The results demonstrated that restorative urban design requires the integration of various environmental factors, including, but not limited to, environmental aesthetics, the nature of ambient acoustics, the presence of specific odors, thermo-environmental factors like solar radiation and wind speed, etc. This research helps practitioners both from academia and industry to draw a better understanding of the current state of research and provide empirical guidance for incorporating health benefits into the design process of public urban spaces.

However, it still has some limitations. The data of this paper were derived from the Web of Science (WoS), and a further combination of literature from multiple databases based on disciplinary distribution is necessary. In addition, the review predominantly includes studies from highly urbanized countries, which may lead to certain cultural and environmental biases. Nevertheless, our study synthesizes the relationships between the spatial properties, sensorial mechanisms, and restoration benefits, providing a novel perspective on how to create urban restorative environments through spatial elements and organization. Furthermore, the multisensory design strategies proposed in this study have broad applicability, offering valuable insights for other cities on how to promote residents' health in the context of increased heat exposure, noise pollution, and a range of urbanization challenges. In the end, potential directions for the future development of relevant research are outlined: (1) Since variables in urban outdoor spaces are inconsistent, such as thermal climate, noise, and air pollution, restoration studies through time could also be further kept in focus. (2) Comparative studies on restorative experiences based on multisensory influences across different types of cities and cultural groups could be conducted. (3) As a product of a multidisciplinary intersection, most of the research only investigates the presence of certain spatial elements on restorative effectiveness and lacks attention to the compound spatial organization (i.e., how spatial elements are constituted together). Meanwhile, current multisensory research has fewer studies on the mechanism of three or more sensory interactions evoked by urban compositions and configurations. With the development of VR, AR, and advanced perceptual sensors, it is recommended to model multisensory experiences determined by practical spatial compositions in order to develop more applicable interventions, which could be implemented into real design and planning practices.

Author Contributions: Conceptualization, W.D. and M.L.; methodology, W.D.; software, W.D. and D.D.; validation, W.D., D.D. and P.S.; formal analysis, W.D.; investigation, W.D. and M.L.; resources, W.D. and M.L.; data curation, W.D.; writing—original draft preparation, W.D.; writing—review and editing, D.D., P.S., R.Z. and M.L.; visualization, W.D. and M.L.; supervision, D.D. and M.L.; project administration, P.S.; funding acquisition, D.D. and M.L. All authors have read and agreed to the published version of the manuscript.

Funding: This research was funded by the National Natural Science Foundation of China (No. 52078159), the Stable Support Programme for Higher Education by Shenzhen R&D Funds (No. GXWD20220811163309001), Young Scientists Fund of the National Natural Science Foundation of China (Grant No. 52208054), and Guangdong Basic and Applied Basic Research Foundation (No. 2023A1515011360).

Data Availability Statement: Data are contained within the article.

Conflicts of Interest: The authors declare no conflicts of interest.

References

1. UN DESA. World Urbanisation Prospects 2018. UN Department of Economic and Social Affairs: Population Dynamics. Available online: https://population.un.org/wup/ (accessed on 16 December 2019).
2. Nagendra, H.; Bai, X.; Brondizio, E.S.; Lwasa, S. The urban south and the predicament of global sustainability. *Nat. Sustain.* **2018**, *1*, 341–349. [CrossRef]
3. Alleyne, G.; Binagwaho, A.; Haines, A.; Jahan, S.; Nugent, R.; Rojhani, A.; Stuckler, D. Embedding non-communicable diseases in the post-2015 development agenda. *Lancet* **2013**, *381*, 566–574. [CrossRef] [PubMed]
4. Berto, R. Exposure to restorative environments helps restore attentional capacity. *J. Environ. Psychol.* **2005**, *25*, 249–259. [CrossRef]
5. Grahn, P.; Stigsdotter, U.A. Landscape planning and stress. *Urban For. Urban Green.* **2003**, *2*, 1–18. [CrossRef]
6. Pazhouhanfar, M.; Kamal, M. Effect of predictors of visual preference as characteristics of urban natural landscapes in increasing perceived restorative potential. *Urban For. Urban Green.* **2014**, *13*, 145–151. [CrossRef]
7. Sluiter, J.K.; Frings-Dresen, M.H.W.; Meijman, T.F.; van der Beek, A.J. Reactivity and recovery from different types of work measured by catecholamines and cortisol: A systematic literature overview. *Occup. Environ. Med.* **2000**, *57*, 298–315. [CrossRef]
8. Joye, Y.; van den Berg, A.E. Restorative environments. In *Environmental Psychology*; Steg, L., van den Berg, A., de Groot, J.I.M., Eds.; Blackwell: East Grinstead, UK, 2013; pp. 57–66.
9. Bratman, G.N.; Hamilton, J.P.; Daily, G.C. The impacts of nature experience on human cognitive function and mental health. Landsc. *Urban Plan.* **2015**, *138*, 41–50.
10. Wang, R.; Feng, Z.; Pearce, J. Neighbourhood greenspace quantity, quality and socioeconomic inequalities in mental health. *Cities* **2022**, *129*, 103815. [CrossRef]
11. Wu, Y.; Liu, Q.; Hang, T.; Yang, Y.; Wang, Y.; Cao, L. Integrating restorative perception into urban street planning: A framework using street view images, deep learning, and space syntax. *Cities* **2024**, *147*, 104791. [CrossRef]
12. Labib, S.M.; Lindley, S.; Huck, J.J. Spatial dimensions of the influence of urban green-blue spaces on human health: A systematic review. *Environ. Res.* **2020**, *180*, 108869. [CrossRef]
13. Mears, M.; Brindley, P.; Jorgensen, A.; Ersoy, E.; Maheswaran, R. Greenspace spatial characteristics and human health in an urban environment: An epidemiological study using landscape metrics in Sheffield, UK. *Ecol. Indic.* **2019**, *106*, 105464. [CrossRef]
14. Cui, L.; Shi, J. Urbanization and its environmental effects in Shanghai, China. *Urban Clim.* **2012**, *2*, 1–15. [CrossRef]
15. Deng, X.; Cao, Q.; Wang, L.; Wang, W.; Wang, S.; Wang, S.; Wang, L. Characterizing urban densification and quantifying its effects on urban thermal environments and human thermal comfort. *Landsc. Urban Plan.* **2023**, *237*, 104803. [CrossRef]
16. Subiza-Pérez, M.; Vozmediano, L.; Juan, C.S. Welcome to your plaza: Assessing the restorative potential of urban squares through survey and objective evaluation methods. *Cities* **2020**, *100*, 102461. [CrossRef]
17. Gavrilidis, A.A.; Niță, M.R.; Onose, D.A.; Badiu, D.L.; Năstase, I.I. Methodological framework for urban sprawl control through sustainable planning of urban green infrastructure. *Ecol. Indic.* **2019**, *96*, 67–78. [CrossRef]
18. Liu, S.; Wang, X. Reexamine the value of urban pocket parks under the impact of the COVID-19. *Urban For. Urban Green.* **2021**, *64*, 127294. [CrossRef]
19. Xu, W.; Wang, H.; Su, H.; Sullivan, W.C.; Lin, G.; Pryor, M.; Jiang, B. Impacts of sights and sounds on anxiety relief in the high-density city. *Landsc. Urban Plan.* **2024**, *241*, 104927. [CrossRef]
20. Lindal, P.J.; Hartig, T. Effects of urban street vegetation on judgments of restoration likelihood. *Urban For. Urban Green.* **2015**, *14*, 200–209. [CrossRef]
21. Subiza-Pérez, M.; Korpela, K.; Pasanen, T. Still not that bad for the grey city: A field study on the restorative effects of built open urban places. *Cities* **2021**, *111*, 103081. [CrossRef]

22. Zhao, J.; Wu, J.; Wang, H. Characteristics of urban streets in relation to perceived restorativeness. *J. Expo. Sci. Environ. Epidemiol.* **2020**, *30*, 309–319. [CrossRef]
23. Karmanov, D.; Hamel, R. Assessing the restorative potential of contemporary urban environment(s): Beyond the nature versus urban dichotomy. *Landsc. Urban Plan.* **2008**, *86*, 115–125. [CrossRef]
24. Reece, R.; Bornioli, A.; Bray, I.; Alford, C. Exposure to Green and Historic Urban Environments and Mental Well-Being: Results from EEG and Psychometric Outcome Measures. *Int. J. Environ. Res. Public Health* **2022**, *19*, 13052. [CrossRef] [PubMed]
25. Lindal, P.J.; Hartig, T. Architectural variation, building height, and the restorative quality of urban residential streetscapes. *J. Environ. Psychol.* **2013**, *33*, 26–36. [CrossRef]
26. Subiza-Pérez, M.; Vozmediano, L.; Juan, C.S. A systematic social observation tool to measure the restorative potential of urban settings/Diseño de una herramienta de observación social sistemática del potencial restaurador de espacios urbanos. *PsyEcology* **2019**, *10*, 257–286. [CrossRef]
27. Zhang, T.; Liu, J.; Li, H. Restorative Effects of Multi-Sensory Perception in Urban Green Space: A Case Study of Urban Park in Guangzhou, China. *Int. J. Environ. Res. Public Health* **2019**, *16*, 4943. [CrossRef]
28. Finlay, J.; Franke, T.; McKay, H.; Sims-Gould, J. Therapeutic landscapes and wellbeing in later life: Impacts of blue and green spaces for older adults. *Health Place* **2015**, *34*, 97–106. [CrossRef]
29. Nghiem, T.; Wong, K.; Jeevanandam, L.; Chang, C.; Tan, L.; Goh, Y.; Carrasco, L. Biodiverse urban forests, happy people: Experimental evidence linking perceived biodiversity, restoration, and emotional wellbeing. *Urban For. Urban Green.* **2021**, *59*, 127030. [CrossRef]
30. Velarde, M.; Fry, G.; Tveit, M. Health effects of viewing landscapes—Landscape types in environmental psychology. *Urban For. Urban Green.* **2007**, *6*, 199–212. [CrossRef]
31. Jiang, B.; Xu, W.; Ji, W.; Kim, G.; Pryor, M.; Sullivan, W.C. Impacts of nature and built acoustic-visual environments on human's multidimensional mood states: A cross-continent experiment. *J. Environ. Psychol.* **2021**, *77*, 101659. [CrossRef]
32. Benita, F.; Tunçer, B. Exploring the effect of urban features and immediate environment on body responses. *Urban For. Urban Green.* **2019**, *43*, 126365. [CrossRef]
33. Elsadek, M.; Liu, B.; Lian, Z.; Xie, J. The influence of urban roadside trees and their physical environment on stress relief measures: A field experiment in Shanghai. *Urban For. Urban Green.* **2019**, *42*, 51–60. [CrossRef]
34. Osborne, T. Restorative and Afflicting Qualities of the Microspace Encounter: Psychophysiological Reactions to the Spaces of the City. *Ann. Assoc. Am. Geogr.* **2022**, *112*, 1461–1483. [CrossRef]
35. Ratcliffe, E.; Gatersleben, B.; Sowden, P.T. Bird sounds and their contributions to perceived attention restoration and stress recovery. *J. Environ. Psychol.* **2013**, *36*, 221–228. [CrossRef]
36. Ha, J.; Kim, H.J. The restorative effects of campus landscape biodiversity: Assessing visual and auditory perceptions among university students. *Urban For. Urban Green.* **2021**, *64*, 127259. [CrossRef]
37. Jiang, L.; Masullo, M.; Maffei, L. Effect of odour on multisensory environmental evaluations of road traffic. *Environ. Impact Assess. Rev.* **2016**, *60*, 126–133. [CrossRef]
38. Lin, I.Y. Evaluating a servicescape: The effect of cognition and emotion. *Int. J. Hosp. Manag.* **2004**, *23*, 163–178. [CrossRef]
39. Schreuder, E.; van Erp, J.; Toet, A.; Kallen, V.L. Emotional Responses to Multisensory Environmental Stimuli: A Conceptual Framework and Literature Review. *Sage Open* **2016**, *6*, 1–19. [CrossRef]
40. Banerjee, T. The Future of Public Space: Beyond Invented Streets and Reinvented Places. *J. Am. Plan. Assoc.* **2001**, *67*, 9–24. [CrossRef]
41. Walters, D. *Designing Community*, 1st ed.; Routledge: London, UK, 2007.
42. Gibson, J.J. *The Ecological Approach to Visual Perception*; Houghton Mifflin: Boston, MA, USA, 1979.
43. Kaplan, R.S.-W.; Kaplan, S. *The Experience of Nature: A Psychological Perspective*; Cambridge University Press: New York, NY, USA, 1989.
44. Guo, X.; Liu, J.; Albert, C.; Hong, X.C. Audio-visual interaction and visitor characteristics affect perceived soundscape restorativeness: Case study in five parks in China. *Urban For. Urban Green* **2022**, *77*, 127738. [CrossRef]
45. Guo, X.; Liu, J.; Chen, Z.; Hong, X.-C. Harmonious Degree of Sound Sources Influencing Visiting Experience in Kulangsu Scenic Area, China. *Forests* **2023**, *14*, 138. [CrossRef]
46. Kaplan, S. The restorative benefits of nature: Toward an integrative framework. *J. Environ. Psychol.* **1995**, *15*, 169–182. [CrossRef]
47. Ulrich, R.S. Aesthetic and Affective Response to Natural Environment. In *Behavior and the Natural Environment*; Altman, I., Wohlwill, J.F., Eds.; Springer: Boston, MA, USA, 1983; pp. 85–125.
48. Ulrich, R.S.; Simons, R.F.; Losito, B.D.; Fiorito, E.; Miles, M.A.; Zelson, M. Stress recovery during exposure to natural and urban environments. *J. Environ. Psychol.* **1991**, *11*, 201–230. [CrossRef]
49. Jeon, J.Y.; Jo, H.I.; Lee, K. Psycho-physiological restoration with audio-visual interactions through virtual reality simulations of soundscape and landscape experiences in urban, waterfront, and green environments. *Sustain. Cities Soc.* **2023**, *99*, 104929. [CrossRef]
50. Lyu, K.; Brambilla, A.; Globa, A.; de Dear, R. A socio-cultural perspective to semi-outdoor thermal experience and restorative benefits—Comparison between Chinese and Australian cultural groups. *Build. Environ.* **2023**, *243*, 110622. [CrossRef]
51. Liu, J.; Kang, J.; Behm, H.; Luo, T. Effects of landscape on soundscape perception: Soundwalks in city parks. *Landsc. Urban Plan.* **2014**, *123*, 30–40. [CrossRef]

52. Dai, T.; Zheng, X. Understanding how multi-sensory spatial experience influences atmosphere, affective city image and behavioural intention. *Environ. Impact Assess. Rev.* **2021**, *89*, 106595. [CrossRef]
53. Chan, S.H.M.; Qiu, L.; Esposito, G.; Mai, K.P. Vertical greenery buffers against stress: Evidence from psychophysiological responses in virtual reality. *Landsc. Urban Plan.* **2021**, *213*, 104127. [CrossRef]
54. Pachimsawat, P.; Tangprasert, K.; Jantaratnotai, N. The calming effect of roasted coffee aroma in patients undergoing dental procedures. *Sci. Rep.* **2021**, *11*, 1384. [CrossRef]
55. Ba, M.; Kang, J. Effect of a fragrant tree on the perception of traffic noise. *Build. Environ.* **2019**, *156*, 147–155. [CrossRef]
56. Xiong, X.; Jin, H.; Hu, W.; Zeng, C.; Huang, Q.; Cui, X.; Zhang, M.; Jin, Y. Benefits of Jasminum polyanthum's natural aromas on human emotions and moods. *Urban For. Urban Green.* **2023**, *86*, 128010. [CrossRef]
57. Kuper, R. Restorative potential, fascination, and extent for designed digital landscape models. *Urban For. Urban Green.* **2017**, *28*, 118–130. [CrossRef]
58. Liu, M.; Nijhuis, S. Mapping landscape spaces: Methods for understanding spatial-visual characteristics in landscape design. *Environ. Impact Assess. Rev.* **2020**, *82*, 106376. [CrossRef]
59. Torku, A.; Chan, A.P.; Yung, E.H.; Seo, J. The influence of urban visuospatial configuration on older adults' stress: A wearable physiological-perceived stress sensing and data mining based-approach. *Build. Environ.* **2021**, *206*, 108298. [CrossRef]
60. Appleton, J. *The Experience of Landscape*; John Wiley and Sons: London, UK, 1975.
61. Cai, M.; Xiang, L.; Ng, E. How does the visual environment influence pedestrian physiological stress? Evidence from high-density cities using ambulatory technology and spatial machine learning. *Sustain. Cities Soc.* **2023**, *96*, 104695. [CrossRef]
62. Tabrizian, P.; Baran, P.K.; Van Berkel, D.; Mitasova, H.; Meentemeyer, R. Modeling restorative potential of urban environments by coupling viewscape analysis of lidar data with experiments in immersive virtual environments. *Landsc. Urban Plan.* **2020**, *195*, 103704. [CrossRef]
63. He, J.; Hao, Z.; Li, L.; Ye, T.; Sun, B.; Wu, R.; Pei, N. Sniff the urban park: Unveiling odor features and landscape effect on smellscape in Guangzhou, China. *Urban For. Urban Green.* **2022**, *78*, 127764. [CrossRef]
64. Van Renterghem, T. Towards explaining the positive effect of vegetation on the perception of environmental noise. *Urban For. Urban Green.* **2019**, *40*, 133–144. [CrossRef]
65. Banerjee, S.; Yik, S.K.; Dzyuban, Y.; Crank, P.J.; Yi, R.P.X.; Chow, W.T. Analysing impacts of urban morphological variables and density on outdoor microclimate for tropical cities: A review and a framework proposal for future research directions. *Build. Environ.* **2022**, *225*, 109646. [CrossRef]
66. Li, J.; Liu, N. The perception, optimization strategies and prospects of outdoor thermal comfort in China: A review. *Build. Environ.* **2020**, *170*, 106614. [CrossRef]
67. Chen, T.; Pan, H.; Lu, M.; Hang, J.; Lam, C.K.C.; Yuan, C.; Pearlmutter, D. Effects of tree plantings and aspect ratios on pedestrian visual and thermal comfort using scaled outdoor experiments. *Sci. Total Environ.* **2021**, *801*, 149527. [CrossRef]
68. Qi, Y.; Chen, Q.; Lin, F.; Liu, Q.; Zhang, X.; Guo, J.; Qiu, L.; Gao, T. Comparative study on birdsong and its multi-sensory combinational effects on physio-psychological restoration. *J. Environ. Psychol.* **2022**, *83*, 101879. [CrossRef]
69. Lam, C.K.C.; Yang, H.; Yang, X.; Liu, J.; Ou, C.; Cui, S.; Kong, X.; Hang, J. Cross-modal effects of thermal and visual conditions on outdoor thermal and visual comfort perception. *Build. Environ.* **2020**, *186*, 107297. [CrossRef]
70. Schweiker, M.; Ampatzi, E.; Andargie, M.S.; Andersen, R.K.; Azar, E.; Barthelmes, V.M.; Berger, C.; Bourikas, L.; Carlucci, S.; Chinazzo, G.; et al. Review of multi-domain approaches to indoor environmental perception and behaviour. *Build. Environ.* **2020**, *176*, 106804. [CrossRef]
71. Torresin, S.; Pernigotto, G.; Cappelletti, F.; Gasparella, A. Combined effects of environmental factors on human perception and objective performance: A review of experimental laboratory works. *Indoor Air* **2018**, *28*, 525–538. [CrossRef]
72. Zhou, Y.H.; Wang, Y.F.; Jin, H.X. Study on the Restorative Benefits of Campus Green Space under Visual and Olfactory Perceptions. *Class. Chin. Gard.* **2023**, *39*, 36–41. [CrossRef]
73. Lee, K.; Wylie, B.; Williams, N.S.; Johnson, K.A.; Sargent, L.D.; Williams, K.J. "It's a little soap opera of its own": Fascinating green roofs offer complexity, movement, sensory engagement, and vast vistas. *Landsc. Urban Plan.* **2024**, *242*, 104925. [CrossRef]
74. Ba, M.; Kang, J. A laboratory study of the sound-odour interaction in urban environments. *Build. Environ.* **2019**, *147*, 314–326. [CrossRef]
75. Zhang, Z.; Měchurová, K.; Resch, B.; Amegbor, P.; Sabel, C.E. Assessing the association between overcrowding and human physiological stress response in different urban contexts: A case study in Salzburg, Austria. *Int. J. Health Geogr.* **2023**, *22*, 15. [CrossRef]
76. Jeon, J.Y.; Jo, H.I.; Lee, K. Potential restorative effects of urban soundscapes: Personality traits, temperament, and perceptions of VR urban environments. *Landsc. Urban Plan.* **2021**, *214*, 104188. [CrossRef]
77. Liu, F.; Liu, P.; Kang, J.; Meng, Q.; Wu, Y.; Yang, D. Relationships between landscape characteristics and the restorative quality of soundscapes in urban blue spaces. *Appl. Acoust.* **2022**, *189*, 108600. [CrossRef]
78. Gallou, E.; Uzzell, D.; Sofaer, J. Perceived place qualities, restorative effects and self-reported wellbeing benefits of visits to heritage sites: Empirical evidence from a visitor survey in England. *Wellbeing Space Soc.* **2022**, *3*, 100106. [CrossRef]
79. Ruotolo, F.; Rapuano, M.; Masullo, M.; Maffei, L.; Ruggiero, G.; Iachini, T. Well-being and multisensory urban parks at different ages: The role of interoception and audiovisual perception. *J. Environ. Psychol.* **2024**, *93*, 102219. [CrossRef]

80. Chung, W.K.; Lin, M.; Chau, C.K.; Masullo, M.; Pascale, A.; Leung, T.M.; Xu, M. On the study of the psychological effects of blocked views on dwellers in high dense urban environments. *Landsc. Urban Plan.* **2022**, *221*, 104379. [CrossRef]
81. Chang, J.; Du, M.; Hong, B.; Qu, H.; Chen, H. Effects of thermal-olfactory interactions on emotional changes in urban outdoor environments. *Build. Environ.* **2023**, *232*, 110049. [CrossRef]
82. Pitt, H. Therapeutic experiences of community gardens: Putting flow in its place. *Health Place* **2014**, *27*, 84–91. [CrossRef]
83. Ba, M.; Li, Z.; Kang, J. The multisensory environmental evaluations of sound and odour in urban public open spaces. *Environ. Plan. B Urban Anal. City Sci.* **2022**, *50*, 1759–1774. [CrossRef]
84. Schnell, I.; Potchter, O.; Yaakov, Y.; Epstein, Y. Human exposure to environmental health concern by types of urban environment: The case of Tel Aviv. *Environ. Pollut.* **2016**, *208*, 58–65. [CrossRef]
85. Lyu, K.; de Dear, R.; Brambilla, A.; Globa, A. Restorative benefits of semi-outdoor environments at the workplace: Does the thermal realm matter? *Build. Environ.* **2022**, *222*, 109355. [CrossRef]
86. Masullo, M.; Maffei, L.; Pascale, A.; Senese, V.P.; De Stefano, S.; Chau, C.K. Effects of Evocative Audio-Visual Installations on the Restorativeness in Urban Parks. *Sustainability* **2021**, *13*, 8328. [CrossRef]
87. Wang, R.; Zhao, J. Effects of evergreen trees on landscape preference and perceived restorativeness across seasons. *Landsc. Res.* **2020**, *45*, 649–661. [CrossRef]
88. Ho, R.; Au, W.T. Effect of Street Performance (Busking) on the Environmental Perception of Public Space. *Front. Psychol.* **2021**, *12*, 647863. [CrossRef] [PubMed]
89. Zhou, P.; Grady, S.C.; Rosenberg, M.W. Creating therapeutic spaces for the public: Elderly exercisers as leaders in urban China. *Urban Geogr.* **2023**, *44*, 59–82. [CrossRef]
90. Henshaw, V. *The Role of Smell in Urban Design*; The University of Salford: Salford, UK, 2011.
91. Cui, X.; Jin, H.X.; Zeng, C.C. A Study on the Influence of Campus Green Space Auditory and Olfactory Interactive Perception on Stress Recovery of College Students. *Class. Chin. Gard.* **2023**, *39*, 26–31. [CrossRef]
92. Zhang, T.; Su, M.; Hong, B.; Wang, C.; Li, K. Interaction of emotional regulation and outdoor thermal perception: A pilot study in a cold region of China. *Build. Environ.* **2021**, *198*, 107870. [CrossRef]
93. Fisher, J.C.; Irvine, K.N.; Bicknell, J.E.; Hayes, W.M.; Fernandes, D.; Mistry, J.; Davies, Z.G. Perceived biodiversity, sound, naturalness and safety enhance the restorative quality and wellbeing benefits of green and blue space in a neotropical city. *Sci. Total Environ.* **2021**, *755*, 143095. [CrossRef]
94. Shi, J.; Huang, J.; Guo, M.; Tian, L.; Wang, J.; Wong, T.W.; Webster, C.; Leung, G.M.; Ni, M.Y. Contributions of residential traffic noise to depression and mental wellbeing in Hong Kong: A prospective cohort study. *Environ. Pollut.* **2023**, *338*, 122641. [CrossRef]
95. Song, I.; Baek, K.; Kim, C.; Song, C. Effects of nature sounds on the attention and physiological and psychological relaxation. *Urban For. Urban Green.* **2023**, *86*, 127987. [CrossRef]
96. Lin, J.; Wang, X.; Lin, G. Performance and atmosphere in urban public spaces: Street music in Guangzhou, China. *Geogr. Res. -Aust.* **2024**, *62*, 279–292. [CrossRef]
97. Narimani, N.; Karimi, A.; Brown, R.D. Effects of street orientation and tree species thermal comfort within urban canyons in a hot, dry climate. *Ecol. Inform.* **2022**, *69*, 101671. [CrossRef]
98. Zhang, T.; Hong, B.; Su, X.; Li, Y.; Song, L. Effects of tree seasonal characteristics on thermal-visual perception and thermal comfort. *Build. Environ.* **2022**, *212*, 108793. [CrossRef]
99. Bell, S.; Hickman, C.; Houghton, F. From therapeutic landscape to therapeutic 'sensescape' experiences with nature? A scoping review. *Wellbeing Space Soc.* **2023**, *4*, 100126. [CrossRef]
100. Qin, Q.; Xuan, W. Multisource Data–Based Urban Odor Mapping and Construction of a Scent-Emotional System for Public Spaces. *J. Urban Plan. Dev.* **2023**, *149*, 05023040. [CrossRef]
101. Ramzy, N.S. Biophilic qualities of historical architecture: In quest of the timeless terminologies of 'life' in architectural expression. *Sustain. Cities Soc.* **2015**, *15*, 42–56. [CrossRef]
102. Luo, S.; Shi, J.; Lu, T.; Furuya, K. Sit down and rest: Use of virtual reality to evaluate preferences and mental restoration in urban park pavilions. *Landsc. Urban Plan.* **2022**, *220*, 104336. [CrossRef]
103. Yakınlar, N.; Akpınar, A. How perceived sensory dimensions of urban green spaces are associated with adults' perceived restoration, stress, and mental health? *Urban For. Urban Green.* **2022**, *72*, 127572. [CrossRef]
104. Jo, H.I.; Jeon, J.Y. Urban soundscape categorization based on individual recognition, perception, and assessment of sound environments. *Landsc. Urban Plan.* **2021**, *216*, 104241. [CrossRef]
105. Zhang, B.; Kang, J. Effect of environmental contexts pertaining to different sound sources on the mood states. *Build. Environ.* **2022**, *207*, 108456. [CrossRef]
106. Zhu, X.; Zhang, Y.; Luo, Y.Y.; Zhao, W. Natural or artificial? Exploring perceived restoration potential of community parks in Winter city. *Urban For. Urban Green.* **2023**, *79*, 127808. [CrossRef]
107. Jiang, X.; Hu, Y.; Larsen, L.; Chang, C.-Y.; Sullivan, W.C. Impacts of urban green infrastructure on attentional functioning: Insights from an fMRI study. *Front. Psychol.* **2023**, *14*, 1047993. [CrossRef]
108. Schebella, M.F.; Weber, D.; Schultz, L.; Weinstein, P. The Nature of Reality: Human Stress Recovery during Exposure to Biodiverse, Multisensory Virtual Environments. *Int. J. Environ. Res. Public Health* **2020**, *17*, 56. [CrossRef]
109. Abdulkarim, D.; Nasar, J.L. Are livable elements also restorative? *J. Environ. Psychol.* **2014**, *38*, 29–38. [CrossRef]

110. Poppe, L.; Van Dyck, D.; De Keyser, E.; Van Puyvelde, A.; Veitch, J.; Deforche, B. The impact of renewal of an urban park in Belgium on park use, park-based physical activity, and social interaction: A natural experiment. *Cities* **2023**, *140*, 104428. [CrossRef]
111. Liu, M.; Mattson, R.; Kim, E. Influences of lavender fragrance and cut flower arrangements on cognitive performance. *Int. J. Aromather.* **2004**, *14*, 169–174. [CrossRef]
112. Bolin, K.; Nilsson, M.E.; Khan, S. The Potential of Natural Sounds to Mask Wind Turbine Noise. *Acta Acust. United Acust.* **2010**, *96*, 131–137. [CrossRef]
113. Nilsson, M.E.; Alvarsson, J.; Rådsten-Ekman, M.; Bolin, K. Auditory masking of wanted and unwanted sounds in a city park. *Noise Control. Eng. J.* **2010**, *58*, 524–531. [CrossRef]
114. You, J.; Lee, P.J.; Jeon, J.Y. Evaluating water sounds to improve the soundscape of urban areas affected by traffic noise. *Noise Control Eng. J.* **2010**, *58*, 477. [CrossRef]
115. Hedblom, M.; Gunnarsson, B.; Schaefer, M.; Knez, I.; Thorsson, P.; Lundström, J.N. Sounds of Nature in the City: No Evidence of Bird Song Improving Stress Recovery. *Int. J. Environ. Res. Public Health* **2019**, *16*, 1390. [CrossRef]
116. Jeon, J.Y.; Lee, P.J.; You, J.; Kang, J. Perceptual assessment of quality of urban soundscapes with combined noise sources and water sounds. *J. Acoust. Soc. Am.* **2010**, *127*, 1357–1366. [CrossRef]
117. Dzyuban, Y.; Hondula, D.; Vanos, J.; Middel, A.; Coseo, P.; Kuras, E.; Redman, C. Evidence of alliesthesia during a neighborhood thermal walk in a hot and dry city. *Sci. Total Environ.* **2022**, *834*, 155294. [CrossRef]
118. Liu, S.; Nazarian, N.; Hart, M.A.; Niu, J.; Xie, Y.; de Dear, R. Dynamic thermal pleasure in outdoor environments—Temporal alliesthesia. *Sci. Total Environ.* **2021**, *771*, 144910. [CrossRef]

land

MDPI

Article

Wisdom of Landscape Construction of China's West Lakes in Historical Period and Its Implications

Lyuyuan Jia, Qing Lin *, Xiyue Wang *, Wenzhen Jia, Ying Zhao, Zhiqing Zhang, Ziqi Cui, Song Chen and Aibo Jin

School of Landscape Architecture, Beijing Forestry University, Beijing 100083, China; lypea0_0@bjfu.edu.cn (L.J.); shibei1273@bjfu.edu.cn (W.J.); zhaoying_la@bjfu.edu.cn (Y.Z.)
* Correspondence: linqing@bjfu.edu.cn (Q.L.); wangxy402@bjfu.edu.cn (X.W.)

Abstract: West Lake holds a significant position internationally. This article studies the spatial distribution, historical development, functions, and the lake–city spatial relationship of 81 West Lakes (WLs), as well as their landscape characteristics. The results indicate the following: (1) The overall spatial distribution of WLs is related to China's history of water conservancy development. (2) The evolution of and functional changes in WLs are influenced by multiple factors such as politics, economy, and culture during different historical periods and are directly related to the will of local administrators. (3) The initial functions of WLs can be categorized into three types, primarily related to urban infrastructure. (4) In terms of spatial relationships, there are four types of spatial relationships between WLs and their water sources and three types of spatial relationships between WLs and cities, forming a common pattern of "Mountains/Hills(–Water)–WL-(–Water)–Cities(–Water, River, Sea)" or "WL(–Water)–Cities(–Water, River, Sea)". (5) The scenery of WLs comprises six elements, including natural basements, water conservancy facilities, human settlements, secularization, landscape architecture, and animal and human activities, all imbued with poetic cultural connotations. Furthermore, this study summarizes three causes of WLs' scenery; excavates the historical wisdom of WLs in terms of a holistic approach, ecological techniques, dynamic management, and landscape esthetics; and subsequently proposes recommendations for lake ecological governance and landscape construction.

Keywords: historic landscape; urban lakes; West Lake; spatial relationship; historical wisdom

1. Introduction

Water is the foundation upon which humanity relies for survival and development. In an era before the widespread rise of industrialization and urbanization, people around the globe generally tended to develop their livelihoods in accordance with the natural environment. Throughout history, lakes, which are widespread on the Earth's surface, played a vital role as environmental supports, showcasing their unique functional value in the harmonious coexistence between humans and nature. A notable example is the Tonle Sap Lake-centered water regulation and storage network developed during the Khmer Empire of Angkor (802–1432), which not only sustained diverse aquatic livelihood activities [1] but also exemplified the adaptive management of the lake's natural hydrological processes [2]. Similar lake-based livelihood systems can be found in Lake Iliamna [3] in the United States, Lake Victoria [4] along the South Australian Victorian border, and Lake Fuquene [5] in

Colombia. In China, Hangzhou's West Lake represents another exemplary case. By integrating with surrounding mountains and waterways, it has historically formed a regional water regulation system that supported the area's livable development and ultimately evolved into a globally renowned cultural landscape [6]. These lakes, through prolonged human–land–water interactions, have fostered models of harmonious coexistence between humans and the environment, embedding ecological wisdom and sustainable development practices within their systems.

However, the dual pressures of accelerating climate change and urbanization have brought the ecological challenges of urban lakes to the forefront [7–10] while also compromising their landscape quality to varying degrees [11,12]. In response to the growing societal demand for high-quality living environments, a global wave of urban lake landscape development has emerged. Yet, the construction of many artificial lakes frequently overlooks the holistic consideration of surrounding water systems. Characterized by small surface areas, shallow depths, and slow water flow [13], such lakes often exhibit poor self-purification capabilities and high susceptibility to external disturbances. Consequently, they frequently confront challenges such as water pollution, diminished ecological value [11], low-quality recreational spaces, and exorbitant management costs. Therefore, constructing urban lakes that are ecological and beautiful and possess certain water conservancy functions while consistently preserving their diverse values poses a common challenge for cities globally. Against this backdrop, scholars have begun to seek answers from history [14].

China has diverse land types and abundant natural mountains and rivers. Under the influence of ancient agricultural civilization and the idea of "harmony between man and nature" [15], the city mode of the organic combination of human settlements and mountains and rivers has been formed [16]. The original meaning of "west lake" is "a lake located on the west side of an ancient city (inside or outside the city)" [17]. China's overall terrain characteristic of "high in the west and low in the east", combined with the traditional urban planning ideology that integrates natural landscapes, has led to the existence of many lakes named "West Lake" (which is translated into "WL" uniformly by the following text) in various cities across China, with a long historical background. Historical records, ancient books, local chronicles, poems, and even folk stories frequently document the genesis, location, scales, functions, management methods, scenery, and human activities associated with WLs. The official Ming Dynasty compilation, *Yongle Encyclopedia*, specifically collated and recorded 36 WLs, while the *Complete Library in the Four Treasuries* of the Qianlong period included over 70 WLs. Additionally, there are some specialized documents on WLs such as *Chronicles of West Lake, Records of West Lake, Chronicles of Huizhou West Lake*, etc. This demonstrates that during China's historical period, WLs were not only widespread but also recognized by both officials and civilians. In traditional urban landscape systems, WLs were often featured prominently in the "Eight Scenic Views" associated with various cities. A total of 33 WLs are included in the cultural system of China's traditional "Eight Scenic Views" of cities [18]. Furthermore, in the "Eight Scenic Views" of cities such as Kunming, Yingzhou, Jinan, etc., there is more than one scenic spot associated with WLs. Additionally, there are scenic spots such as the "Ten Scenic Views" of Hanghzou WL, "Eight Scenic Views" of Huizhou WL, and "Eight Scenic Views" of Fuzhou WL, indicating that WLs have become a universal scenic symbol in the context of China's specific natural geography and sociocultural context [19,20]. However, the accelerating processes of globalization and urbanization have gradually eroded many traditional cultural landscapes [21,22], including WLs. This has led to a diminishing understanding and appreciation of their historical and cultural value. The inscription of Hangzhou WL Cultural Landscape on the UNESCO World Heritage List in 2011 revived global interest in traditional Chinese lake scenery exemplified by WLs [6], and the "WL Model" has gradually become an ideal paradigm for creating livable environments [23]. Therefore, systematically reviewing WLs in China's history; understanding their formation and development mechanisms; uncovering the wisdom behind their construction and the historical, social, scenic, and cultural values they embody; and summarizing the traditional landscape creation characteristics of this type of scenery can help enhance the local and international recognition of WLs. This can not only ensure the scientific and effective implementation of cultural heritage protection in national spatial planning but also provide experience for the construction of modern urban lakes.

The primary disciplines that initially paid attention to regional history were history, sociology, and archeology. By the 1980s, landscape anthropology emerged, focusing on the interactive relationship between humans and their surrounding environment [24]. However, such research largely depends on in-depth fieldwork and interviews [25,26], tends to focus on specific regions, and lacks applicability to historical studies that span multiple epochs. In late 20th century, Berkes' [27] concept of Traditional Ecological Knowledge (TEK) gained widespread recognition and was extensively applied in various fields such as regional cultural studies, traditional revitalization, and biodiversity conservation [28]. Similar terminology includes indigenous ecological knowledge (IEK) [29], local ecological knowledge and local knowledge [30], etc. However, these studies have primarily focused on the aspects of local technology, local society, and local culture, with applications mostly in the research of local settlements, traditional livelihoods, and ethnic settlements. There has been relatively little attention paid to the spatial features of the landscape. The research methods of landscape history originated from the work of historical regionalists, specialists in environmental issues, or landscape archeologists [31,32] and traditional research methods including fieldwork and desk research. In the mid- to late 20th century, advances in Geographic Information Systems (GISs) catalyzed a spatial turn in landscape research, enhancing methodological precision and enabling more visual analyses [33]. In the early 21st century, the methodology of landscape historiography was introduced into the study of traditional landscapes, and many scholars employed this approach to explore the characteristics of traditional landscapes influenced by specific individuals [34,35], geographical units [36], or particular types of feature elements [37,38].

Current research on WLs is quite extensive. Zhang [39] restored the scenery of Xuzhou WL during the Tang and Song dynasties using historical documents. Yang [35] integrated landscape historical research methods to interpret the heritage value of Yangzhou WL. Zhang [6] studied the evaluation system of Hangzhou WL at the time of its inclusion in the World Heritage List and compared it with the actual perceptions of Chinese tourists, interpreting the cultural landscape of Hangzhou WL through this comparison. Liang [40] analyzed the "Eight Scenic Views" and "Twelve Scenic Views" of Huizhou WL in the Ming and Qing dynasties, exploring its landscape and cultural characteristics. Yan [41] conducted a comparative study of Hangzhou WL, Yingzhou WL, and Huizhou WL, all influenced by Su Shi, and summarized the "Dongpo Model" and its ecological principles. While these studies offer valuable insights, their research subjects are either case studies of one WL or comparative studies of three to four WLs within similar regions. The research content mainly focuses on three aspects: first, historical description; second, the analysis of the characters and literary works related to the construction of the WLs' landscape [42,43]; and third, the current landscape characteristics and value of the WLs [44–46]. A few national studies, such as that by Mao who analyzed 36 WLs recorded in *Yongle Encyclopedia·Auspicious Water*, identify common characteristics of these WLs [20] and proposed the ecological wisdom embedded in "WL Culture" [21]. Wu [17], under the guidance of Professor Mao, and Qiu [18], advised by Professor Li, surveyed over 80 WLs across China. However, these studies focusing on WLs across China primarily emphasize data statistics and the cultural phenomena they present. The systematic summarization of the spatial and landscape characteristics of these WLs remains inadequate, highlighting a gap in current research.

To address the gaps in previous research, this paper adopts a landscape historiography approach to systematically study WLs during the historical period prior to 1911 AD. This study focuses on exploring the spatial distribution, historical development, initial functions, and water source characteristics of WLs. By summarizing the spatial features and landscape elements of WLs, this research seeks to uncover the underlying patterns and regularities behind the complex phenomena across different times and spaces. This paper aims to answer several core questions: How were these WLs formed and developed historically? What are the landscape features of WLs? What are the causes and connotations of the WLs' scenery? What wisdoms can be drawn from the landscape construction of WLs in the historical period? Through a systematic review and analysis, this study strives to enhance the recognition of WLs both domestically and internationally. It also aims to provide theoretical insights and practical guidance for the design and construction of modern urban lakes, drawing from the historical wisdom embedded in the lake-scape.

2. Study Object and Methodology

2.1. Study Object

Spatially, this study takes all WLs clearly recorded as "West Lake" in historical records within China as the study object (Figure 1). Temporally, the "historical period" referred to in this study pertains to the time span prior to 1911AD. The rationale for delineating this temporal scope is grounded in the fact that during the period before 1911AD, productivity was relatively low, and people primarily relied on natural conditions for their livelihoods, accumulating a wealth of TEK. Following 1919AD, modern industries emerged in China, towns expanded gradually, and numerous large-scale water conservancy facilities were newly constructed, resulting in significant alterations to water and soil resources across the national territory.

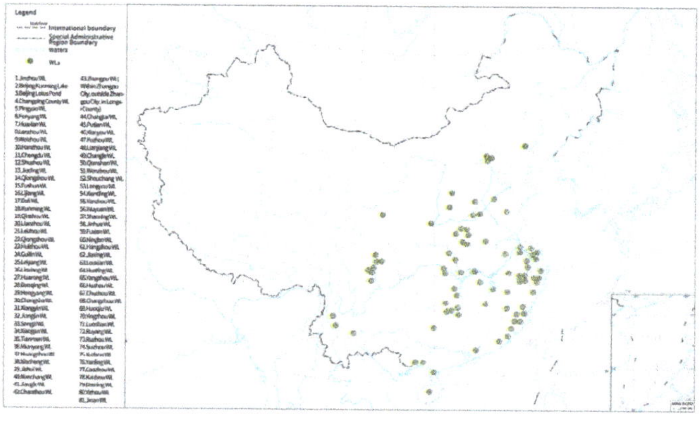

Figure 1. Study object (source: this Chinese map was downloaded from the "Standard Map Service System of the Ministry of Natural Resources", with an approval number of GS(2019)4345. The spatial locations of WLs are self-drawn by the author).

2.2. Data Source

The types of data used in this study include historical documents, historical maps, historical surveying and mapping data, satellite images, photographs, and Chinese standard maps. Specifically, the standard map of China was downloaded from the "Standard Map Service System of the Ministry of Natural Resources (http://211.159.153.75/index.html (accessed on 8 December 2024))". Other historical data and their specific sources are detailed in Table 1.

Table 1. Ancient records of WLs (source: self-drawn).

Type	Name	Dynasty	Author
History Book	*History of the Han Dynasty*	Han (76–84)	Ban Gu
	Old Book of Tang	Five Dynasties (941–945)	Liu Xu, et al.
	New Book of Tang	Song (1044–1060)	Ouyang Xiu, Song Qi, et al.
	Records of Dreams of Glory	Song (1274)	Wu Zimu
Book by Category	*Yongle Encyclopedia*	Ming (1408)	Anonymity
a Collection of Books	*the Complete Library in the Four Treasuries*	Qing (1792)	Yong Rong, Ji Yun, etc.
Monograph	*Classic of Water Resources*	Northern Wei of the Northern Dynasties (386–534)	Li Daoyuan
	Literally Essence of Historical Geography or Essentials of Geography for Reading History	Qing (1875–1908)	Gu Zuyu
	Annals of Three Mountains in the Chunxi Years	Qing (1182)	Liang Kejia
	Records of West Lake	Qing	Lin Dachuan
	Chronicles of West Lake	Qing (1735)	Fu Wanglu, etc.
	Chronicles of Fuzhou West Lake	1916	He Zhenbai
	Mountain records of West Lake	1924	Rao E
	Chronicles of Huizhou West Lake	1947	Zhang Youren
Local Chronicle	*Linan prefectural chronicle*	Song (1250)	Zhou Chong
	Yanzhou prefectural chronicle	Ming (1573–1620)	Yang Shouren
	Kaizhou prefectural chronicle	Ming (1522–1566)	Wang Chongqing
	Suizhou prefectural chronicle	Ming (1368–1644)	Li Mengyang
	Yanling prefectural chronicle	Qing (1659)	He Elian
	Wanping prefectural chronicle	Qing (1662–1722)	Anonymity
	Changzhou prefectural chronicle	Qing (1662–1722)	Yu Kunxiu
	Jiaxing prefectural chronicle	Qing (1685)	He Zhi
	Huizhou prefectural chronicle	Qing (1688)	Lyu Yingkui
	Ruyang prefectural chronicle	Qing (1690)	Qiu Tianying
	Songzi prefectural chronicle	Qing (1696)	Chen Lin
	Chengwu prefectural chronicle	Qing (1702)	Zhao Cijin
	XInghufu Putian prefectural chronicle	Qing (1705)	Jin Gaoxie
	Xuzhou prefectural chronicle	Qing (1745)	Zhen Ruzhou
	Huangzhou prefectural chronicle	Qing (1749)	Wang Qing
	Changle prefectural chronicle	Qing (1763)	He Shijun
	Jiangle prefectural chronicle	Qing (1765)	Li Yongxi
	Nanchang prefectural chronicle	Qing (1789)	Chen Lansen, Wang Wenyong
	Huating prefectural chronicle	Qing (1791)	Feng Dinggao, Li Tingjing
	Shaoxing prefectural chronicle	Qing (1792)	Li Hengte
	Jiangling prefectural chronicle	Qing (1794)	Cui Longjian, Wei Yao
	Lianjiang prefectural chronicle	Qing (1805)	Li Beng
	Songjiang prefectural chronicle	Qing (1818)	Song Rulin
	Xiangyin prefectural chronicle	Qing (1823)	Yan Zhaolang
	Qinzhou prefectural chronicle	Qing (1834)	Zhu Chunnian

Table 1. *Cont.*

Type	Name	Dynasty	Author
Local Chronicle	*Qiongshan prefectural chronicle*	Qing (1857)	Li Wenxuan
	Wenzhou prefectural chronicle	Qing (1865)	Li Wan
	Huoqiu prefectural chronicle	Qing (1870)	Lu Dingxiao, Wang Yinqing
	Linxiang prefectural chronicle	Qing (1872)	En Rong
	Qianshan prefectural chronicle	Qing (1873)	Zhang Yanheng
	Huarong prefectural chronicle	Qing (1882)	Sun Bingyu
	Xiaogan prefectural chronicle	Qing (1883)	Zhu Xibai
	Leshan prefectural chronicle	Qing (1887)	Gong Chuanfu
	Jinxian prefectural chronicle	Qing (1875–1908)	Dai Mei
	Luoshan prefectural chronicle	Qing	Ge Quan
	Yunnan prefectural chronicle	1912–1949	Zhou Jifeng
	Tianmen prefectural chronicle	1922	Hu Yi
	Xuchang prefectural chronicle	1924	Zhang Shaoxun
	Ruzhou prefectural chronicle	1963	Cheng Tiangui
Poetry and Literature	*A poem written by Su Shi to a good friend about the dredged West Lake*	Song (1092)	Su Shi
	Wen-Gong poetry and literature collection	Song (1066)	Hu Su
	Mengliang Record	late Southern Song	Wu Zimu
	Old Affairs of Wulin	late Song to early Yuan	Zhou Mi

2.3. Research Methodology

This article applied research methodologies from landscape history, specifically including the following 4 key steps (Figure 2): (1) Define the Study Object. This study identified 81 WLs that were clearly recorded as "West Lake" within China in historical records. (2) Build an Information Database. Combining historical research and fieldwork investigations, we completed their geographic location labeling in GIS using ArcGIS Pro 3.0 software. Additionally, information such as the names, excavation types, water source types, initial functions, city–WL spatial relationshipswas collected to establish a historical information database (for details, see Appendix A). (3) Categories, Descriptions, and Graphical Analysis. Based on quantitative data, qualitative descriptions were provided for different results, and morphological methods were employed for classification and illustration. (4) Propose Landscape Strategies. The historical landscape formation of WLs was unearthed, the historical wisdom in their construction was deciphered, and based on this, lake-scape construction strategies that meet the needs of modern cities were proposed.

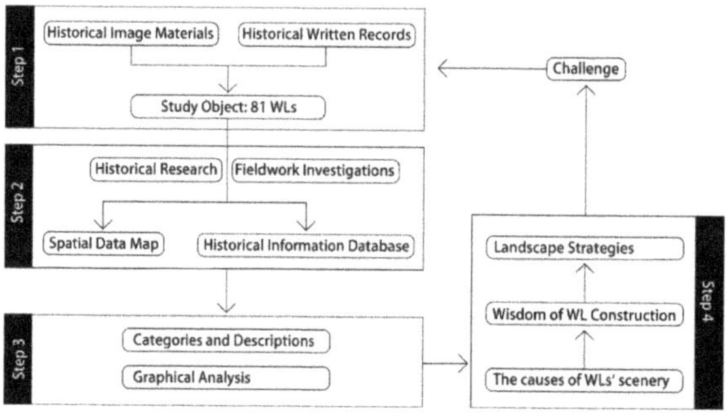

Figure 2. Research framework (source: self-drawn).

3. Results

3.1. Spatial Distribution of WLs

We superimposed the spatial distribution of the 81 WLs on ten major cultural zones in China and found that, in terms of geographical distribution, WLs are mainly concentrated in Jiangnan Water Town Cultural Zone, where the number of WLs (38) is nearly one-half of the country's. The North China Plain Cultural Zone (15) and the South China Mazu Cultural Zone (13) are also relatively well distributed, followed by the Sichuan Basin Cultural Zone (7), and there are four and three WLs in the Loess Plateau Cultural Zone and the Yunnan–Guizhou Plateau Cultural Zone, respectively, and one is located in the Northeast Black Earth Cultural Zone (Figure 3a). There are 66 WLs concentrated in the eastern belt of China's geographical divisions (Figure 3b). Almost all of the WLs are located in the southeastern half of the area divided by the Hu Line, with only the Lanzhou WL located northwest of this line (Figure 3c). In the figure titled "WLs and the National-Level River Basin", it is clearly observable that WLs are predominantly distributed within the Yangtze River Basin (34), followed by the Southeastern River Basin (17). The number of WLs in the Huaihe River Basin (8), the Pearl River Basin (7), and the Yellow River Basin (7) is roughly equivalent. Additionally, there are 4 WLs located in the Haihe River Basin, 3 in the Southwest River Basin, and 1 in the Songliao River Basin (Figure 3d). Spatially, we can see that the distribution of these historical WLs is closely related to China's history of water management and the development trend of water conservancy [47,48].

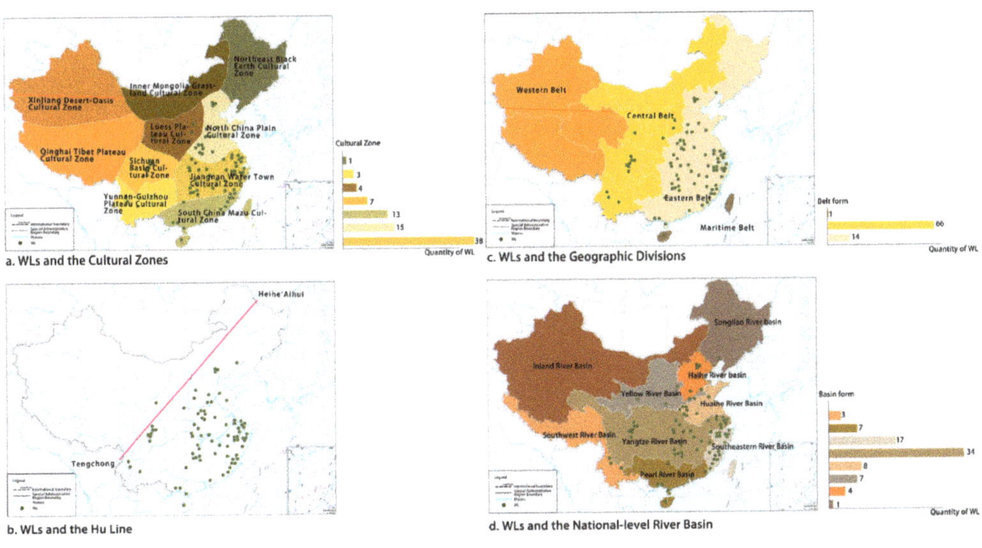

Figure 3. Spatial distribution of WLs (source: the Chinese map was downloaded from the "Standard Map Service System of the Ministry of Natural Resources", with an approval number of GS(2019)4345. The rest were self-drawn by the author).

3.2. Historical Development of WLs

As early as the Han and Wei dynasties, historical records concerning WLs began to emerge. In the book *the History of the Han Dynasty · Geography*, Hangzhou WL was recorded during the Eastern Han Dynasty. Hangzhou WL and the Beijing Lotus Pond were recorded in the book *Commentary on the Waterways Classic* written by Li Daoyuan in the Northern Wei Dynasty. Prior to the prosperous Tang Dynasty (618 AD), WLs were primarily used as city infrastructure, with a simple and rustic appearance. A few lakes with recreational functions were mainly visited by local officials [46]. During this time, the number of WLs was increasing, albeit at a slow rate. From the Tang Dynasty to the Northern Song Dynasty, the social environment was stable, and cultural endeavors flourished, leading to the rise and prosperity of public gardens [49,50]. WLs gradually became a place of public attraction. However, by the Northern Song Dynasty, factors such as conflict, the expansion of residential land use, and the lack of lake management and maintenance led to the siltation, drying up, and disappearance of WLs from the Northern Song period to the end of the Qing Dynasty. Corresponding the historical development of WLs to the periods of Chinese history, we can see that the rise and fall of WLs and their landscape construction were related to the social, economic, and cultural environment of the time. During periods of political stability, economic prosperity, and cultural flourishing, more construction of WLs occurred. This construction was directly influenced by the will of local managers, as evidenced by cases such as Beijing's Kunming Lake and Hangzhou WL, which is inseparable from the management system of feudal society in Chinese history.

3.3. *Initial Function of WLs*

Based on their original functions, WLs can be divided into 3 types: WLs that were developed for water conservancy, WLs that were developed for city security defense, and WLs that were developed under the influence of Feng Shui or secular enlightenment.

3.3.1. WLs That Were Developed for Water Conservancy

This type of WL was often developed from low-lying water bodies within or outside the city. Beijing Kunming Lake, for example, was initially a natural pond formed by the convergence of spring water and rainfall from the Yuquan and Xishan areas in the western suburbs of Beijing, serving as a natural lake for flood detention and water storage in the outskirts. During the Tianbao era of the Tang Dynasty (AD 744), Lu Nanjin, the county magistrate of Fu County, led the people in constructing dikes at Fuxian WL. According to historical records, Fuxian WL was capable of irrigating as much as 16.67 km^2 [1] of farmland [24,25]. This type of WLs fulfills the functions of flood storage and regulation, as well as water supply, thereby ensuring water safety and meeting the water demands of cities and surrounding agricultural areas.

3.3.2. WLs That Were Developed for City Security Defense

This type of WL can be subdivided into 2 subclasses. One subclass concerns the construction of the city to take soil; this type of WL mostly consists of artificial lakes. On the one hand, they can prevent the impact of external water on the city; on the other hand, they can store water for future use. Xuzhou WL is adjacent to Xuchang County; after taking soil to build the ancient city wall, the low-lying area where the soil was taken was filled with water to form Xuzhou WL. The other subclass consists of WLs that serve as a moat for cities. This type of WL, in conjunction with the surrounding mountains/hills, water systems, or walls, collectively formed the defense system of the early city. Wuyuan WL is a part of the moat formed by artificial water diversion. Chaozhou WL constituted the western defense of the ancient Chaozhou city with mountains/hills (Figure 4).

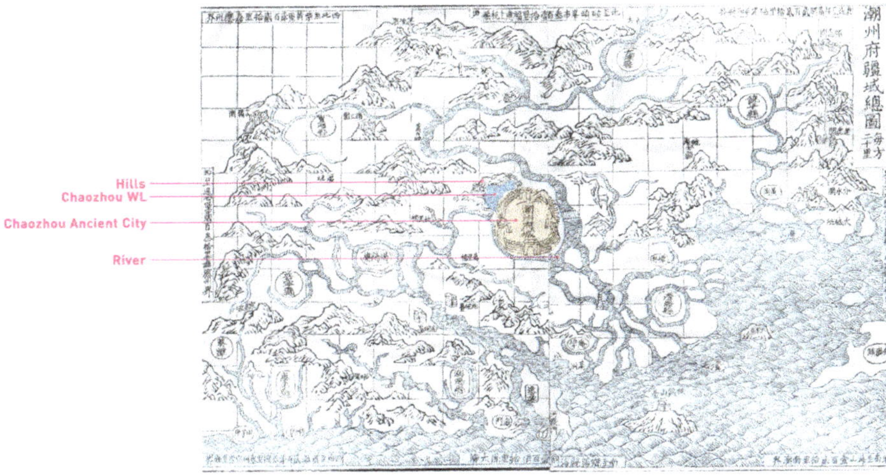

Figure 4. Chaozhou WL and ancient Chaozhou City (source: ancient map of Chaozhou City is sourced from *Chaozhou prefectural chronicle*, and the rest are self-drawn by the author).

3.3.3. WLs That Were Developed Under the Influence of Feng Shui or Secular Enlightenment

This type of WL was predominantly constructed under the influence of Feng Shui principles for city planning, secular enlightenment, and the leisure and entertainment needs of scholars and officials. Most of these WLs originated from natural lakes and served multiple functions, including flood storage and water supply. Specifically, they can be divided into 3 subclasses: religious release WLs, tour ornamental WLs, and Feng Shui WLs. During the Qianyuan period of the Tang Emperor Suzong's reign (758 AD–759 AD), there was a practice of "setting up as many as religious release ponds near the city wall"; at that time, the local official (Yan Zhenqing) dredged Chaozhou WL and turned it into a religious release pond [38]. As tour ornamental WLs, the Beijing Lotus Pond, Yingzhou WL, and Guilin WL are all natural lakes. Yingzhou WL, with its picturesque natural scenery, attracted a local nobleman to build an imperial garden here more than 1000 years ago [51], in the period before the Common Era. During the Tang Dynasty, before the emergence of Shuzhou WL, the site was originally known as "Dongting" (a tourist attraction). By the Northern Song Dynasty, Zhao Yuedao, while diverting water to irrigate farmland, conveniently channeled it to the location of "Dongting," resulting in the formation of Shuzhou WL. The WL, together with the earlier Dongting, created a space for sightseeing [52]. Feng Shui WLs can be exemplified by Zhangpu WL. According to *Chorography of Zhang pu Prefecture*, in 1241 AD, the county magistrate Zheng Shishen deemed the original Feng Shui of the Zhangpu county seat's location unfavorable [53]. Consequently, he ordered the transformation of a portion of farmland on the west side of the county seat into a lake, giving rise to Zhangpu WL.

3.4. *Relationship Between WLs and Water Sources*

According to the existing literature, the excavation type of WLs can be divided into 3 categories: natural lakes, artificial lakes, and natural–artificial composite lakes. Apart from the 13 WLs that lack historical records, it is evident that natural lakes constitute the majority (52.2%). According to the type of water sources of WLs, WLs can be divided into 4 types: ①WLs formed by foothill catchments, ②WLs formed at the confluence of rivers or along the banks of rivers, ③ WLs formed by the combined flow of springs and streams, and ④ WLs formed by artificial diversion (Figure 5; for details, see Appendix A). Among these types of WLs, artificial WLs are all attributed to the ④ water source type, and natural WLs are developed from the lakes formed by mountain water or evolved from streams and bays (Figure 6).

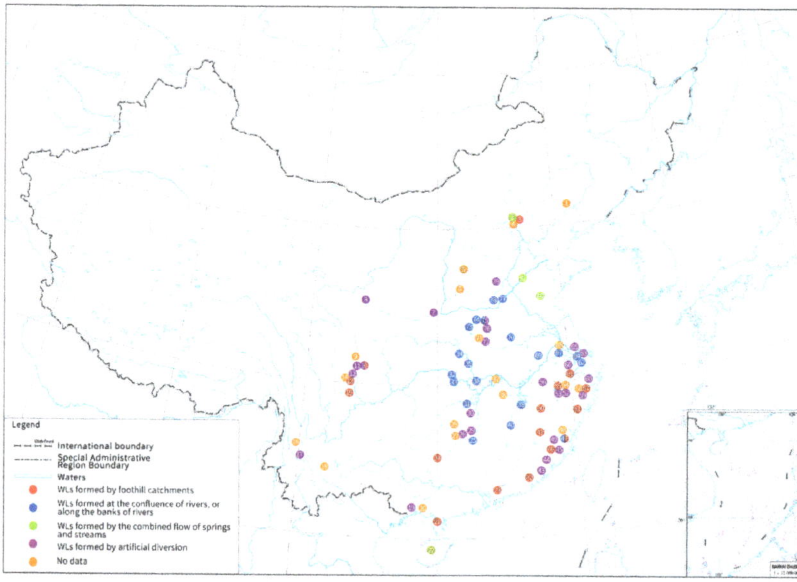

Figure 5. The water source types of the 81 WLs (source: the Chinese map was downloaded from the "Standard Map Service System of the Ministry of Natural Resources", with an approval number of GS(2019)4345. The rest were self-drawn by the author).

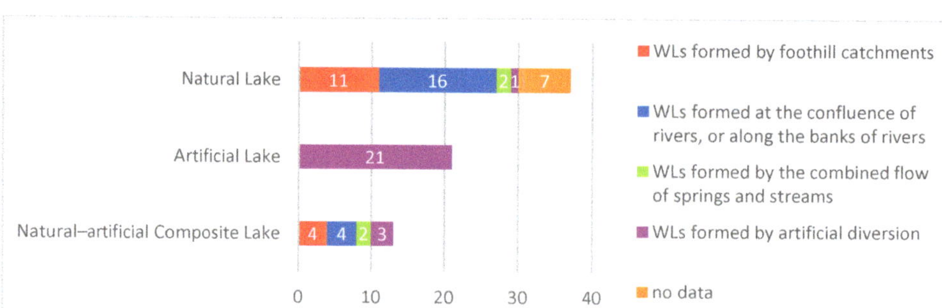

Figure 6. Statistics on the relationship between WLs and water sources (source: self-drawn).

WLs formed by foothill catchments have close relationships with West Mountain/Hills, usually located at the foot of the West Mountain/Hills, with spatial characteristics of "Mountain/Hills–WL" (Figure 7(a-1,a-2)). WLs formed at the confluence of rivers or along the banks of rivers are frequently linked to natural watercourses through canals and irrigation systems, forming a spatial pattern of "River–WL" (Figure 7(b-1,b-2)). WLs formed by the combined flow of springs and mountain streams are mostly situated in the piedmont plain of West Mountain/Hills, with depressions, springs, and ponds often found between WLs and mountains, forming a spatial pattern of "Mountain/Hills–Spring–WL" (Figure 7(c-1,c-2)).

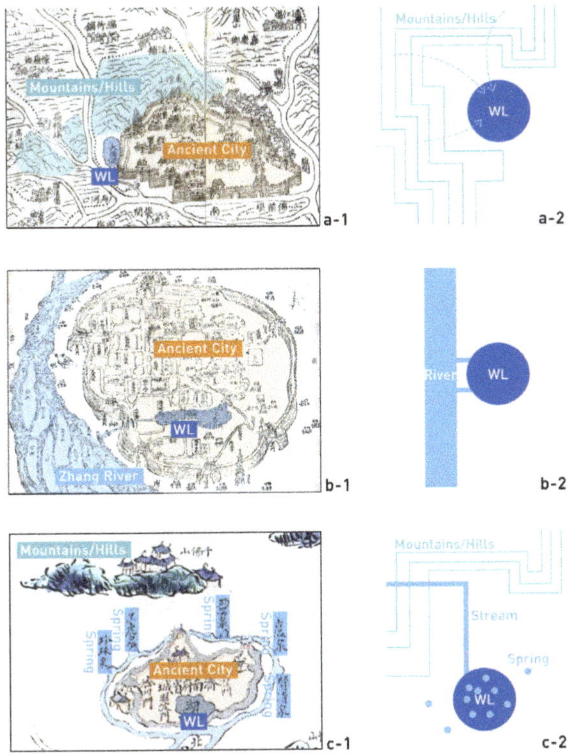

Figure 7. Relationship between WLs and water sources. (**a-1**) Map analysis of WLs formed by foothill catchments (source: ancient map is sourced from *Leshan prefectural chronicle*, and the rest are self-drawn by the author). (**a-2**) Schematic diagrams of WLs formed by foothill catchments (source: self-drawn). (**b-1**) Map analysis of WLs formed at the confluence of rivers or along the banks of rivers (source: ancient map is sourced from *Nanchang prefectural chronicle*, and the rest are self-drawn by the author). (**b-2**) Schematic diagrams of WLs formed at the confluence of rivers or along the banks of rivers (source: self-drawn). (**c-1**) Map analysis of WLs formed by the combined flow of springs and streams (source: ancient map is sourced from *Shandong Yellow River map, respectively*, and the rest are self-drawn by the author). (**c-2**) Schematic diagrams of WLs formed by the combined flow of springs and streams (source: self-drawn).

3.5. Spatial Relationship Between WLs and Cities

Due to the early functions of WLs in urban defense and water supply, there was a close relationship between WLs and their corresponding ancient cities. Based on the positional relationship between cities and WLs, they can be classified into 3 types: WLs that were separated from the city, WLs that were adjacent to the city, and WLs that were embedded in the city (Figure 8; for details, see Appendix A).

WLs that were separated from the city maintain a certain spatial distance from the city and are always connected to the urban water system network through rivers, ditches, and canals. Beijing Kunming Lake, which was located far from the Beijing ancient city, was connected to the lake and water systems within the city via the Zao River, ultimately linking up to the canal. On a larger spatial scale, together with the mountains in the western suburbs, Beijing Kunming Lake formed a scenic landscape pattern of "Mountains/Hills–Spring-WL–River–City–Water (Canal)" (Figure 9).

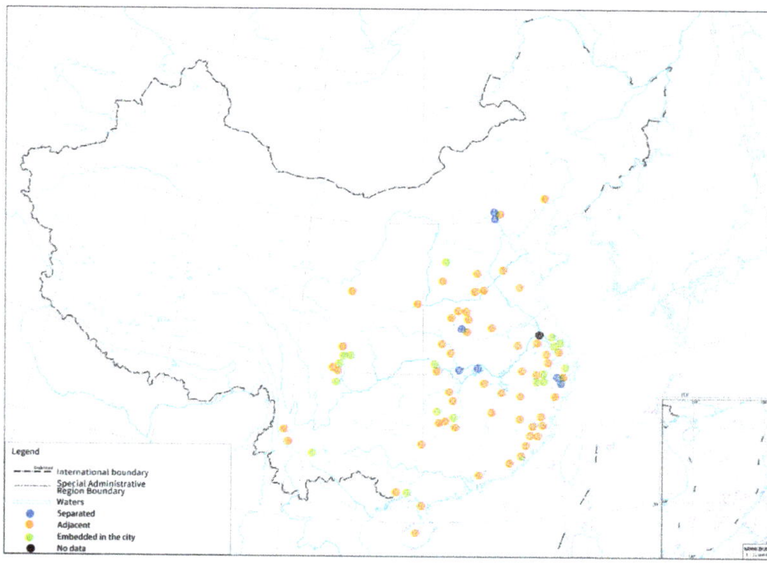

Figure 8. The city–WL spatial relationships of the 81 WLs (source: the Chinese map was downloaded from the "Standard Map Service System of the Ministry of Natural Resources", with an approval number of GS(2019)4345. The rest were self-drawn by the author).

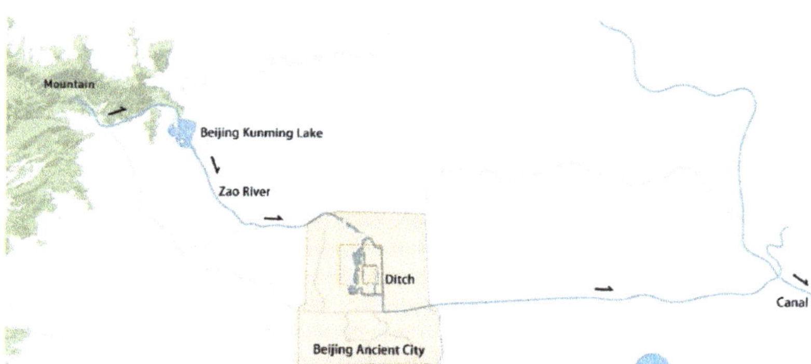

Figure 9. The pattern of WLs that were separated from the city (Beijing ancient city and Beijing Kunming Lake) (source: referring to the article *Research on Traditional Urban Landscape System of Beijing based on Water Conservancy System*, self-drawn).

WLs that were adjacent to the city are closely related to the development of cities. The survival and development of these WLs are often related to the urban population or the size of the city. Because of the important functions of Hangzhou WL, such as flood storage, water supply, agriculture, and fisheries, as well as the landscape, Hangzhou has always had a "city–WL" landscape, "being surrounded by mountains on three sides and adjacent to the city on one side" (Figure 10a). At the end of the Tang Dynasty, with the expansion of Fuzhou city and the construction of interconnected water systems both inside and outside the city, Fuzhou WL, on the one hand, served as a reservoir storage lake to channel water for flood relief and supply water to the city and surrounding fields; on the other hand, it constituted a defense for the city together with the city walls and water system shorelines [54,55]. Despite the continuous expansion and development of the city during the Song, Ming, and Qing dynasties, the scenic pattern of "Mountains/Hills–WL–City–River" has remained consistent and unbroken on a regional scale (Figure 10b).

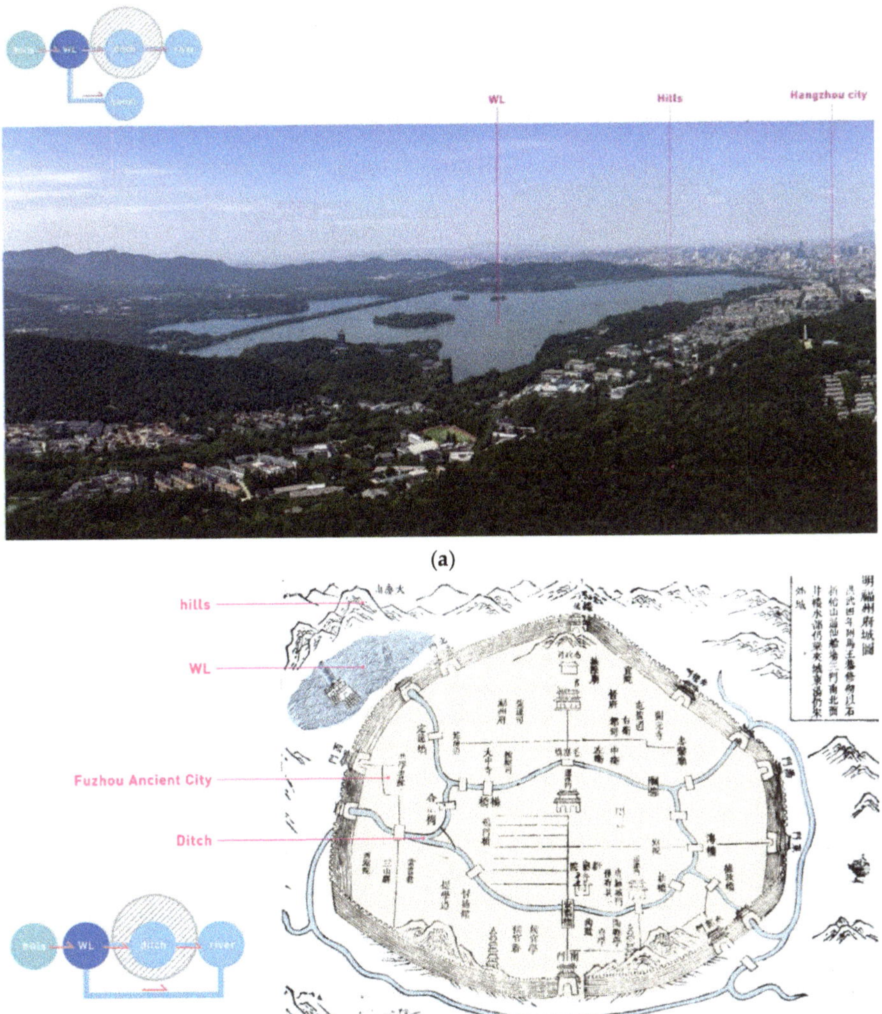

Figure 10. The pattern of WLs that were adjacent to the city. (**a**). Hangzhou's city and lake landscape of "being surrounded by mountains on three sides and adjacent to the city on one side" (photo credits: Hangzhou Landscape and Cultural Relics Bureau). (**b**). The spatial relationship between Fuzhou WL and Fuzhou Ancient City (source: ancient map of Fuzhou City is sourced from *Minzu koan (Fujian province)*, and the rest are self-drawn by the author).

WLs that were embedded in the city have a close interconnection with the city's water system. Ningbo WL, with the function of storing freshwater in the urban water ecosystem, constitutes the river network pattern of "three rivers, six ditches, and one lake nestled in the city center" with other rivers and ponds in the city (Figure 11a). To address the waterlogging issues of Nanchang WL during the Tang Dynasty, inner and outer sluice gates and multiple drainage canals were constructed. These canals connected Nanchang WL with the surrounding river and ditch systems, evolving into a city water system featuring "three lakes and nine ditches" by the Qing Dynasty (Figure 11b). Jinan WL Prefecture boasts a relatively vast water surface, and as the city expanded, it was gradually incorporated within the ancient city. This integration created a unique city–WL landscape for Jinan, often described as "mountains encircle the city and half of the inner space is occupied by Jinan WL" (Figure 11c).

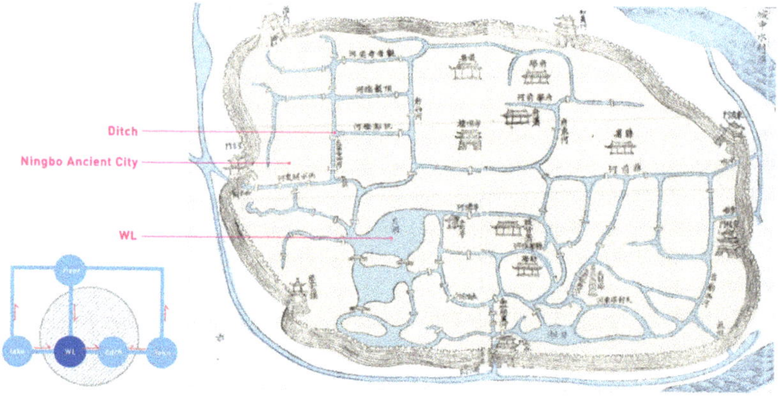

(**a**)

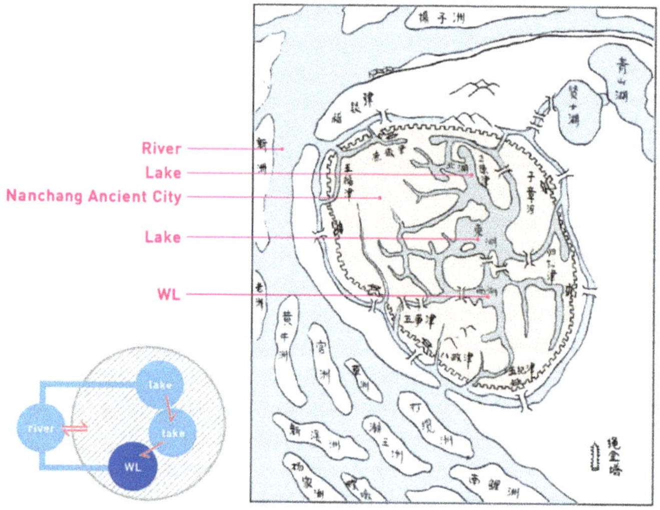

(**b**)

Figure 11. *Cont.*

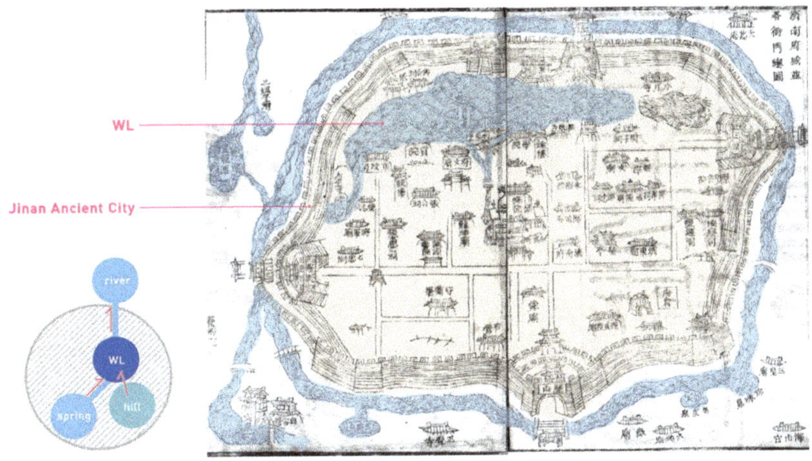

(**c**)

Figure 11. The pattern of WLs that were embedded in the city. (**a**). The spatial relationship between Ningbo WL and Ningbo Ancient City (source: ancient map of Ningbo City is sourced from *Ying County prefectural chronicle*, and the rest are self-drawn by the author). (**b**). The spatial relationship between Nanchang WL and Nanchang Ancient City (source: ancient map of Nanchang City is sourced from *Revisiting the changes and development of ancient Nanchang city*, and the rest are self-drawn by the author). (**c**). The spatial relationship between Jinan WL and Jinan Ancient City during the Qing Dynasty (source: ancient map of Jinan City is sourced from *Jinan prefectural chronicle*, and the rest are self-drawn by the author).

3.6. Landscape Characteristics of WLs

3.6.1. The Scenery of the Region

Based on the relationship between WLs and their water sources, as well as their spatial relationships with cities, it can be observed that at the regional scale, the scenery of WLs was characterized by either "Mountains/Hills(–Water)–WL(–Water)–Cities(–Water, River, Sea)" or "WL(–Water)–Cities(–Water, River, Sea)" (Figure 12).

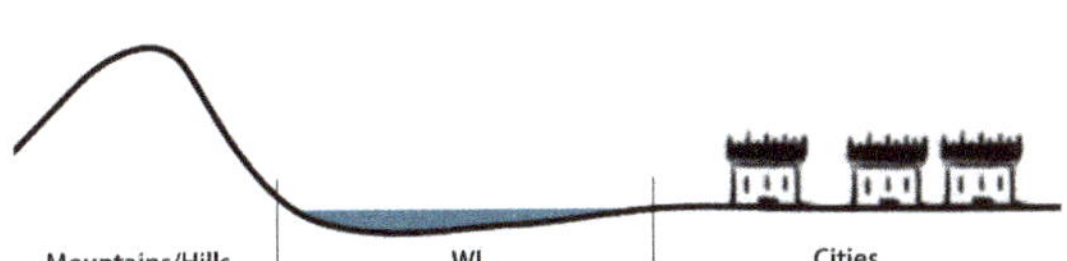

Figure 12. The scenery of WLs at the regional scale (source: self-drawn).

3.6.2. Diversity of Landscape Elements

Numerous historical records show that the scenic elements of WLs include mountains/hills, waterfalls, islands, bridges, towers, temples, and plants, as well as scenes of animal and human activities. These landscape elements can be categorized into 7 types: natural basements, water conservancy facilities, human settlements, secularization, landscape architecture, and animal and human activities (Table 2). Similarly, in the images of WLs (Figure 13), we can also observe the depiction of elements such as mountains/hills, lakes/water systems, bridges, dikes, islands, towers, temples, and villages.

Table 2. Scenic elements of WLs (self-drawn).

Type	Landscape Elements	Examples
Natural basements	mountains/hills, lakes, waterfalls, clouds, shadows, etc.	"Foshan Reflection (佛山倒影)" and "Gathering Waves at Night (汇波晚照)" of Jinan WL, "Ancient Castellated with Slanted Sun (古堞斜阳)" in Fuzhou WL, "Autum Moon over Calm Lake (平湖秋月)" and "Twin Peaks Piercing the Clouds (双峰插云)" of Hangzhou WL, etc.
Water conservancy facilities	dikes, islands, bridges, gates, stone towers, water conservancy monuments, wells, etc.	Six Wells, the Su Causeway in "Spring Dawn on Su Causeway (苏堤春晓)" scenery, the stone tower as a water level warning in the "Three Pools Mirroring the Moon (三潭映月)" scenery, and the causeways and islands in the "Two Causeways and Three Islands (两堤三岛)" scenery in Hangzhou WL. Pinghu Causeway in Huizhou WL, Ou Causeway and Su Causeway in Yingzhou WL, etc.
Human settlements	ancient cities, walls, mansions, houses, taverns, fairs, villages, farmlands, etc.	"Rice Field Fragrance (绿稻香来) " and "Fair Trade (竹楼小市)" in Yangzhou WL. The mansion of Zhang, Fengle tavern in Hangzhou WL.
Secularization	temples, Taoist abbeys, ancestral halls, academies, pavilions, tomb stone carvings, release pools, etc.	"The Sound of Temple Bells in the Morning (西禅晓钟)" in Fuzhou WL. Kou's memorial Temple, Jing's memorial Temple, Laiquan Academy in Leizhou WL. Ruzi Pavilion (celebrity memorial) in Nanchang WL, etc.
Landscape architecture	pavilions, terraces, towers, boats, lotuses, willows, pines and other garden plants	Pavilions named "to take the wind (乘风亭)", "to wait for the moon (待月亭)", and "gradually into a better scene (渐入佳景亭)" and temple called "drunk guests return to the temple (醉客方归庵)" of Chaozhou WL, "fairy bridge with willow on the side (仙桥柳色)" of Fuzhou WL, "Lotus in the Breeze at Crooked Courtyard (曲院风荷)" and "Listening to Orioles Singing in the Willows (柳浪闻莺)" in Hangzhou WL, etc.
Animal and human activities	chopping firewood, fishing, ferry competitions, fetes, sightseeing, etc.	"When the first rays of dawn break through, fisherman sings in the WL (丰湖渔唱)" and "The woodcutter returned home with the sun descending in the west (半径樵归)" in Huizhou WL. Dragon Boat Festival race of Fuzhou WL [49], etc.

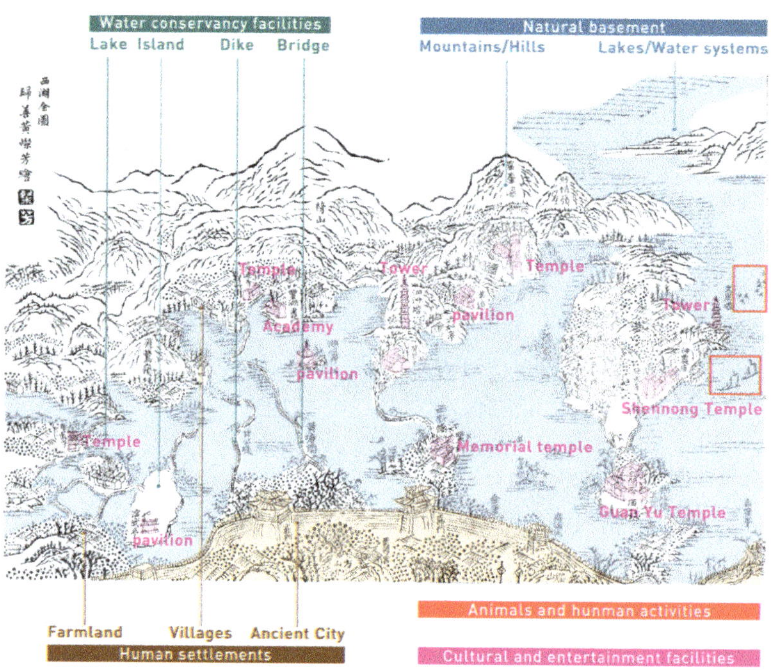

Figure 13. Full picture of Huizhou WL (Source: ancient picture of Huizhou WL is sourced from *Huizhou prefectural chronicle*, and the rest are self-drawn by the author).

3.6.3. Poetic Landscapes with Cultural Connotations

Since ancient times, scholars and patriots such as Su Shi, Ouyang Xiu, and Xu Ruzi have played pivotal roles in shaping WLs. They either participated in their dredging and construction projects, composed poems and essays to capture their breathtaking scenery, or established their residences. These have promoted the landscape construction of WLs and endowed WLs with deep cultural connotation.

Chen Cheng, the chief keeper of Huizhou in the Northern Song Dynasty, vigorously reformed Huizhou WL and proposed the "Eight Scenic Views of Huizhou WL" (Figure 14), which was the first time that WLs were incorporated into the urban landscape system. Later, the "Eight Scenic Views" landscape system with WLs as the landscape element emerged continuously. For example, the "Eight Scenic Views of Licheng" depicts Jinan WL, including "A Misty Rain of Magpies (鹊华烟雨)", "Sink Mave at Night (汇波晚照)", "Boating on the WL (明湖泛舟)", "Autumn Wind Brushes the Surface of the Jinan WL (历下秋风)", as well as the "Night Moon on the Fushun WL (西湖夜月)" in the "Ten Scenic Views of Fushun", "Green Embrace of the Leizou WL (西湖翠拥)" in the "Eight Scenic Views of Leizhou", etc.

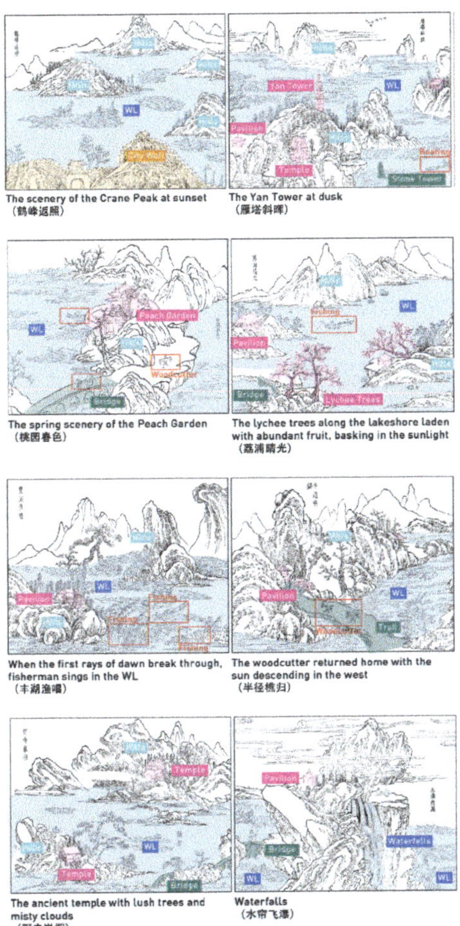

Figure 14. The "Eight Scenic Views Of Huizhou WL" (source: pictures of the "Eight Scenic Views of Huizhou WL" are sourced from *Records of Huizhou West Lake*, and the rest are self-drawn by the author).

The spatial environment of WLs and the cultural support of literati and writers since ancient times have contributed to WLs having many spiritual and cultural treasures being created about them, such as poems, lyrics, literature, paintings, songs, Fugues, writings, couplets, myths and legends, folk stories, etc., which have been handed down to the present day. In return, these artistic and cultural treasures feed back to WLs and have become important carriers that influence the cultural connotations of WLs and spread the cultural heritage of WLs. When applying for World heritage status at Hangzhou WL, someone once wrote the following: "On the broken bridge, my mind seemed to travel to ancient times, while I drank with Su Dongpo and Bai Juyi under the moon." WLs are not only a landscape, a garden, but also a cultural landscape containing history, a kind of thought and life. WLs have become the spiritual sustenance in people's hearts.

4. Discussion

4.1. *The Causes of WLs' Scenery in History*

The natural environment of WLs aligns with the traditional Chinese esthetic of landscape, which has always been favored by literati and poets [56]. Over time, lakes naturally experience sedimentation [57]. To maintain the urban functions of WLs, ancient efforts continuously involved activities such as dredging, water diversion, and the construction of dikes and sluices to sustain the integrity of lakes. The infrastructure formed during these processes began to evolve into scenic elements during the Song Dynasty. Furthermore, the landscape relationship formed by WLs and their surrounding natural elements, including mountains, rivers, and springs, aligns with China's time-honored esthetic appreciation of mountains and waters. It also resonates with the idealized realms revered by religious cultures such as Taoism [58,59], attracting the construction of cultural facilities such as academies, temples, memorials for celebrities, villages and hamlets, and gardens [60]. This beautiful environment also attracts a variety of wildlife, creating ecological scenes like "Orioles Singing in the Willows (柳浪闻莺)", "Apes Howling at Cold Spring (冷泉猿啸)", and "Wild Geese Roosting on Reed Beach (芦汀宿雁)". In summary, the causes of WL scenery can be encapsulated as follows: (1) WLs generally possess a natural foundation that resonates with the traditional Chinese "Mountain–Water" landscape esthetic. (2) Throughout history, the facilities generated during the dredging of WLs have not only supported the functions of livelihood but also sustained the ecological environment of the lake, and these facilities have gradually become part of the scenery. (3) Various leisure facilities have been constructed, drawn by the exquisite natural and cultural scenery of WLs.

4.2. *Historical Wisdom in the Construction of WLs*

4.2.1. Holistic Approach

WLs are an integral part of the regional mountain and water spatial system. Historically, there were no clear professional boundaries among ecology, urban design, and engineering; instead, thinking about problems was holistic. Therefore, in history, when dealing with regional water relations, people often started from an overall perspective of mountains, water, and cities. This holistic thinking not only comprehensively addressed the water use challenges of urban and rural settlements but also maintained the water storage function of freshwater. City defense-type WLs near city walls could also form a natural moat around the city, ensuring the environmental security of many ancient cities in China and serving as a foundation for orderly urban development. Moreover, this connectivity also helps to maintain the water flow and ecological purification of lakes.

4.2.2. Ecological Techniques

Due to the limitations of technological conditions in early times, people often adapted to the natural environment to carry out production activities. Before the Song Dynasty, the construction of WLs was primarily focused on water conservancy and city security defense functions. To meet the needs of the city at that time, continuous water conservancy activities such as dredging, desilting, and water diversion into the lake were conducted. The silt cleared from the lake was used to create facilities like islets, embankments, and dikes, and vegetation was planted along the shores to stabilize the banks. These engineering techniques, initially for water conservancy construction, often also promoted the development of secondary wetlands in WLs, providing habitats and activity corridors for a variety of organisms.

4.2.3. Dynamic Management

Human civilization has arisen from the use of wisdom to compensate for the deficiencies of natural conditions, a concept reflected in the historical development of WLs. Many WLs exhibit characteristics of "swampification", and the famous ones that persist to this day have also experienced periods of desiccation and siltation in their histories, especially from the Northern Song Dynasty to the end of the Qing Dynasty. However, after each occurrence of sedimentation and functional degradation in WLs, people carried out dredging and management in line with the historical appearance of WLs. Furthermore, after each dredging and desilting project, people would incorporate new elements into WLs based on the developmental needs and social contexts of the time, enabling WLs to iterate and renew themselves continuously.

4.2.4. Landscape Esthetics

Based on the ancient human settlement construction ideas of harmonizing with nature and emphasizing the scenic beauty of mountains and waters, the scenic development of WLs has also been extended to the regional perspectives of mountains, rivers, and cities. Furthermore, the imperial examination system for selecting officials since the Sui and Tang dynasties promoted knowledgeable individuals to serve as local officials, thereby advancing the high-level cultural construction of WLs [61]. Coupled with the numerous poems, inscriptions, and ink paintings about WLs left by literati of various dynasties, the scenery of WLs far surpasses that of other lakes, making it a symbol of Chinese landscape esthetics.

4.3. Implications for Modern Lake Construction

Nowadays, the water conservancy functions of WLs, such as flood storage and water supply, have gradually diminished. In contrast, the value of WLs as an urban leisure green space and public sightseeing area has become increasingly prominent. Consequently, the protection and management efforts for WLs often focus solely on maintaining their water bodies while neglecting the connectivity between them and the surrounding mountains/hills and water systems. Furthermore, the perceptions of the WL landscape are predominantly confined to the lake itself. There is a notable lack of consideration for landscape cognition and style preservation from a broader citywide or regional perspective. As a result, some historical aspects of WLs have been compromised or lost entirely. Issues such as improper protection leading to surface shrinkage, deterioration in water quality, decline in ecological function, and loss of scenic beauty, among others, have emerged alongside many newly constructed urban lakes that frequently suffer from poor hydrology and ecological challenges. These problems are not unique to WLs in China; regions with TEK such as the Tonlé Sap Lake in Southeast Asia [28], the Shurkul Lake Landscape in Khorezm [62], and the polder landscapes in the Netherlands all face similar challenges.

In the current era, the preservation of historical facilities necessitates creative reconstruction to adapt to contemporary development needs [63]. However, this process demands a profound understanding and cognition of history; otherwise, it may lead to the misinterpretation of the landscape and a loss of historical continuity [64]. Drawing upon the historical wisdom embodied in the construction of WLs, we can reflect on the following four points:

Firstly, it is hoped that professionals from various fields, including ecology, environmental protection, urban planning, landscape architecture, and water conservancy, will engage in understanding traditional knowledge, integrating natural scientific techniques with social scientific methods organically.

Secondly, when protecting and renewing historical lakes such as WLs, it is crucial to draw inspiration from the ancients' holistic view of the lake's environment from a regional perspective, thereby establishing a systematic and continuous approach to water management.

Thirdly, history has shown that many beautiful spots in WLs also serve certain functions in maintaining production, living, and ecology. Therefore, when constructing lake landscapes, in addition to esthetic and recreational considerations, more thought should be given to how these facilities can serve the needs of today's cities. These lakes, with the advancement of technology, can no longer play the same water conservancy and urban defense functions as in historical times, requiring practitioners in urban planning, landscape architecture, and water conservancy to rethink new development directions for these historical WLs.

Fourthly, landscape construction techniques and the culture of "Eight Scenic Views" in WL construction can continue into the new era of landscape construction. This also reminds us that when constructing landscapes, we can infuse a certain amount of local culture, which can make the landscape more iconic and identifiable.

4.4. Limitations and Future Prospects

This study focuses on internationally significant and regionally distinct WLs, examining their long-term developmental characteristics throughout historical periods to uncover their spatial features and developmental mechanisms, which is particularly crucial for understanding urban lakes. This research adopts the methodology of landscape history, analyzing the historical construction wisdom of WLs through categories and descriptions and graphical analysis. This provides valuable insights for the study of TEK. Additionally, the historical information database of 81 WLs established in this study facilitates further in-depth research on WLs in China.

The sample data involved in this study were obtained through manual identification and may be subject to the omission of sample subjects. In the future, machine learning methods could be introduced to train the extraction of "West Lake" content from ancient local records. In addition, historical records have certain limitations. To ensure the completeness and comprehensiveness of the research data, the historical information database constructed in this study was processed with "no data" for parts where no information was found. In this study, the absence of these data does not affect the conclusions and significance of this research. Similarly, due to the limitations of historical materials, there are parts in this study where the author conducted textual research and inference based on multiple historical sources and modern scholarly studies.

5. Conclusions

The landscape of WLs is dynamic, blending nature and culture. Initially, WLs primarily served as a water conservancy facility or a city security defense, with a simple appearance, and they have been preserved through successive generations of dredging. The development of scholar gardens during the Song Dynasty promoted the gradual formation of WLs into an urban landscape that integrates natural beauty with cultural values. Ancient WLs, as they exist today, embody people's understanding of the natural environment, climate, hydrology, and geography, as well as information about water conservancy projects, social environments, and local customs from different periods and regions, especially in the context of establishing homes and settlements. As a result, WLs contain various landscape elements, including natural basements, water conservancy facilities, human settlements, secularization, and landscape architecture, as well as the activities of animals and people. At the regional scale, WLs form a common landscape pattern of "Mountains/Hills(–Water)–WL(–Water)–Cities(–Water, River, Sea)" or "WL(–Water)–Cities(–Water, River, Sea)". WLs are not only a model of historical Chinese landscapes but also a quintessential example of the symbiosis between cities and water globally. The wisdom in the construction of WLs—its holistic approach, ecological techniques, dynamic management, and landscape esthetics—is a treasure of human history. This knowledge not only helps in understanding and protecting WLs in China but also offers valuable insights for today's lake ecological management and landscape planning.

Author Contributions: Conceptualization, L.J.; Methodology, L.J., X.W., W.J. and S.C.; Software, W.J.; Validation, L.J., W.J., Z.Z. and Z.C.; Formal analysis, L.J., W.J. and Y.Z.; Investigation, L.J., Q.L., X.W., W.J. and Y.Z.; Resources, L.J., Q.L. and X.W.; Data curation, L.J. and Y.Z.; Writing—original draft, L.J.; Writing—review & editing, L.J., Q.L., X.W., Z.Z., S.C. and A.J.; Visualization, L.J. and Y.Z.; Supervision, L.J.; Project administration, L.J., Q.L. and Z.C.; Funding acquisition, Q.L. and X.W. All authors have read and agreed to the published version of the manuscript.

Funding: This research was funded by National key R&D Program of China "Urban Ecological Space Control and Layout Optimization Technology" (No. 2022YFC3800203).

Data Availability Statement: The original contributions presented in this study are included in the Appendix A. Further inquiries can be directed to the corresponding authors.

Conflicts of Interest: The authors declare no conflict of interest.

Appendix A

No.	Name	Excavation Type	Water Source	Initial Function	City–WL Spatial Relationships	Situation
1	Jinzhou WL	Natural Lake	No Data	No Data	adjacent	Existing
2	Beijing Kunming Lake	Natural–artificial Composite Lake	③	□	separated	Existing
3	Beijing Lotus Pond	Natural–artificial Composite Lake	①	■●	adjacent	Existing
4	Changping County WL	No Data	No Data	No Data	separated	Disappeared
5	Pingyao WL	No Data	No Data	No Data	embedded in the city	Disappeared
6	Fenyang WL	Natural Lake	No Data	No Data	adjacent	Existing
7	Huaxian WL	Artificial Lake	④	○	adjacent	Disappeared
8	Lanzhou WL	Artificial Lake	④	○	adjacent	Existing
9	Weizhou WL	No Data	No Data	No Data	adjacent	Disappeared
10	Hanzhou WL	Natural Lake	①④	◆○	embedded in the city	Existing
11	Chengdu WL	Artificial Lake	④	○	embedded in the city	Disappeared
12	Shuzhou WL	Artificial Lake	④	□■●	embedded in the city	Existing
13	Jiading WL	Natural Lake	①	○	adjacent	Disappeared
14	Qiongzhou WL	No Data	No Data	○	adjacent	No Data
15	Fushun WL	Natural Lake	①	□●	embedded in the city	Existing
16	Lijiang WL	Natural Lake	No Data	○	adjacent	Disappeared

No.	Name	Excavation Type	Water Source	Initial Function	City–WL Spatial Relationships	Situation
17	Dali WL	Natural–artificial Composite Lake	④	□	adjacent	Existing
18	Kunming WL	Natural Lake	No Data	○	embedded in the city	Existing
19	Qinzhou WL	Artificial Lake	④	◆	adjacent	Disappeared
20	Lianzhou WL	No Data	No Data	No Data	embedded in the city	Disappeared
21	Leizhou WL	Natural Lake	①	■○	adjacent	Existing
22	Qiongzhou WL	Natural Lake	③	■	adjacent	Existing
23	Huizhou WL	Natural–artificial Composite Lake	①	¤	adjacent	Existing
24	Guilin WL	Natural Lake	①	○●	adjacent	Existing
25	Leiyang WL	Natural Lake	②	□	adjacent	Existing
26	Linxiang WL	Artificial Lake	④	No Data	adjacent	No Data
27	Huarong WL	No Data	No Data	No Data	adjacent	No Data
28	Baoqing WL	Natural Lake	No Data	No Data	embedded in the city	Disappeared
29	Hengyang WL	Natural Lake	②	○	embedded in the city	Existing
30	Changsha WL	Natural–artificial Composite Lake	④	○	adjacent	Existing
31	Xiangyin WL	Natural Lake	②	No Data	adjacent	Existing
32	Jianglin WL	Natural Lake	②	○	embedded in the city	Existing
33	Songzi WL	Natural Lake	②	No Data	adjacent	No Data
34	Xiaogan WL	Natural Lake	②	No Data	adjacent	Disappeared
35	Tianmen WL	Natural Lake	②	□	adjacent	Existing
36	Mianyang WL	Natural Lake	②	No Data	separated	Existing
37	Huangzhou WL	Natural Lake	No Data	○	separated	Existing
38	Xincheng WL	No Data	No Data	No Data	adjacent	No Data
39	Jishui WL	Natural Lake	②	○●	adjacent	Disappeared
40	Nanchang WL	Natural Lake	②	□■	adjacent	Existing
41	Jiangle WL	No Data	①	□○	adjacent	Disappeared
42	Chaozhou WL	Natural–artificial Composite Lake	①	◆○	adjacent	Existing
43	Zhangpu WLs(3)	Artificial Lake	④	¤	two adjacent and one embedded in the city	No Data
44	Changtai WL	Artificial Lake	④	□■	adjacent	Disappeared
45	Putian WL	Artificial Lake	④	No Data	adjacent	Existing
46	Xianyou WL	Natural Lake	①	No Data	adjacent	Disappeared
47	Fuzhou WL	Natural–artificial Composite Lake	①②	□■◆	adjacent	Existing
48	Lianjiang WL	No Data	No Data	No Data	adjacent	Disappeared
49	Changle WL	Artificial Lake	④	□■	adjacent	Disappeared
50	Qianshan WL	Natural Lake	①	○●	adjacent	Existing
51	Wenzhou WL	Natural Lake	①	■	adjacent	Existing
52	Shouchang WL	Artificial Lake	④	■	embedded in the city	Existing
53	Longyou WL	Artificial Lake	④	○	embedded in the city	Disappeared
54	Xinding WL	No Data	No Data	No Data	embedded in the city	Disappeared
55	Yanzhou WL	Natural Lake	①	□■¤	adjacent	Existing
56	Wuyuan WL	Artificial Lake	④	□◆	adjacent	Disappeared
57	Shaoxing WL	Natural Lake	①	No Data	adjacent	Disappeared
58	Jinhua WL	Natural Lake	No Data	No Data	separated	Existing
59	Fuxian WL	Artificial Lake	④	■	separated	Existing
60	Ningbo WL	Artificial Lake	④	■○	embedded in the city	Existing
61	Hangzhou WL	Natural Lake	①	□■	adjacent	Existing
62	Jiaxing WL	Natural Lake	②	○	adjacent	Existing
63	Louxian WL	Artificial Lake	④	○	embedded in the city	Disappeared

No.	Name	Excavation Type	Water Source	Initial Function	City–WL Spatial Relationships	Situation
64	Huating WL	Natural Lake	②	No Data	embedded in the city	Existing
65	Yangzhou WL	Artificial Lake	④	□	embedded in the city	Existing
66	Huzhou WL	Artificial Lake	④	◆	adjacent	Existing
67	Chuzhou WL	Natural Lake	②	○	adjacent	Existing
68	Changzhou WL	No Data	No Data	No Data	No Data	No Data
69	Huoqiu WL	Natural Lake	②	□	adjacent	No Data
70	Yingzhou WL	Natural Lake	②	□○●	adjacent	Existing
71	Luoshan WL	No Data	No Data	■	separated	Disappeared
72	Ruyang WL	Artificial Lake	④	○	adjacent	Existing
73	Ruzhou WL	Natural Lake	②	■	adjacent	Disappeared
74	Suizhou WL	Natural–artificial Composite Lake	②	□	adjacent	Existing
75	Xuzhou WL	Artificial Lake	②④	■◆○	adjacent	Existing
76	Yanling WL	Artificial Lake	④	■○	adjacent	Disappeared
77	Caozhou WL	Natural–artificial Composite Lake	②	○	adjacent	Existing
78	Kaizhou WL	Natural–artificial Composite Lake	②	◆	adjacent	Existing
79	Daming WL	Natural–artificial Composite Lake	④	■	adjacent	Disappeared
80	Yizhou WL	Natural Lake	③	No Data	adjacent	Existing
81	Jinan WL	Natural–artificial Composite Lake	③	■	adjacent	Existing

Note: In the Water Source column, ① represents the WLs formed by foothill catchments; ② represents the WLs formed at the confluence of rivers or along the banks of rivers; ③ represents the WLs formed by the combined flow of springs and streams; and ④ represents the WLs formed by artificial diversion. In the Initial Function column, □ represents WLs that were developed for flood storage, ■ represents WLs that were developed for water supply, ◆ represents WLs that were developed for city security defense, ¤ represents WLs that were developed for religious release, ○ represents WLs that were developed for tour ornamental purposes, and ● represents WLs that were developed for Feng Shui.

Notes

1 The original area recorded is 500 Q in Tang Dynasty units. In Chinese history, 1 Q equals 10/3 hectare or approximately 667 square meters.

References

1. Molyvann, V. *Modern Khmer Cities*; Art Media Resources Ltd.: Chicago, IL, USA, 2003; pp. 3–4.
2. Nuorteva, P.; Keskinen, M.; Olli, V. Water, livelihoods and climate change adaptation in the Tonle Sap Lake area, Cambodia: Learning from the past to understand the future. *J. Water Clim. Chang.* **2010**, *1*, 87–101. [CrossRef]
3. Kugo, Y. Iliamna Lake Nanvarpak ethnogeography: Yup'ik and contemporary place names and stories of these places. *North. Rev.* **2021**, *52*, 5–28. [CrossRef]
4. Lennon, J.; Davis, K. Cultural landscape protection at Lake Victoria, a managed water supply. *Landsc. Res.* **2019**, *45*, 265–279. [CrossRef]
5. Guerrero, P. "Dos siglos de desecación en la Laguna de Fúquene (Colombia): Impactos en la pesca artesanal". *Agua Territ.* **2014**, *4*, 47–58. [CrossRef]
6. Zhang, R.; Taylor, K. Cultural landscape meanings. The case of West Lake, Hangzhou, China. *Landsc. Res.* **2019**, *45*, 164–178. [CrossRef]
7. Gao, H.; Bohn, T.J.; Podest, E.; McDonald, K.C.; Lettenmaier, D.P. On the causes of the shrinking of Lake Chad. *Environ. Res. Lett.* **2011**, *6*, 329–346. [CrossRef]
8. Qin, B.; Xu, P.; Wu, Q.; Luo, L.; Zhang, Y. Environmental Issues of Lake Taihu, China. *Hydrobiologia* **2007**, *581*, 3–14. [CrossRef]
9. Yang, G.; Ma, R.; Zhang, L.; Jiang, J.; Yao, S.; Zhang, M.; Zeng, H. Lake status, major problems and protection strategy in China. *Lake Sci.* **2010**, *22*, 799–810.

10. Wang, J.; Xu, W.; Ma, F.; Zhu, J.; Zhang, S. Thoughts and suggestions on treatment of urban lakes. *Yangtze River* **2022**, *53*, 41–47+71.
11. Chang, N.; Zhang, Q.; Wang, Q.; Luo, L.; Wang, X.C.; Xiong, J.; Han, J. Current status and characteristics of urban landscape lakes in China. *Sci. Total Environ.* **2020**, *712*, 135669. [CrossRef] [PubMed]
12. Su, W.; Gu, C.; Yang, G.; Chen, S.; Zhen, F. Measuring the impact of urban sprawl on natural landscape pattern of the Western Taihu Lake watershed, China. *Landsc. Urban Plan.* **2010**, *95*, 61–67. [CrossRef]
13. Birch, S.; McCaskie, J. Shallow urban lakes: A challenge for lake management. *Hydrobiologia* **1999**, *395*, 365–378. [CrossRef]
14. Hoagland, S.J. Integrating Traditional Ecological Knowledge with Western Science for Optimal Natural Resource Management. *IK Other Ways Knowing* **2017**, *3*, 1–15.
15. Wang, S. The "harmony between man and nature" theory and the ancient Chinese human settlement building. *J. Northwest Univ. (Nat. Sci. Ed.)* **2009**, *39*, 915–920.
16. Wang, S.; Gao, Y.; Li, X. Research on spatial patterns of landscape and humanities in Chinese cities. *Urban Plan. Forum* **2019**, *1*, 27–32.
17. Wu, X. The Historical Procedure and Aesthetical Interpretation of the "West Lake" Culture. Master's Thesis, Chongqing University, Chongqing, China, 2014.
18. Qiu, Y. Research on the "West Lake Landscape Model" of Chinese Traditional City Planning. Ph.D. Thesis, Beijing Forestry University, Beijing, China, 2023.
19. Mao, H.; Du, C.; Chen, X. The ecological wisdom and practical significance of "West Lake Culture". *Landsc. Archit.* **2014**, *6*, 59–63.
20. Mao, H. The evolving procedure and historical significance of the West Lake—A statistic analysis on literatures about the West Lake in The Yongle Canon: Six Pattern Lake. *Chin. Landsc. Archit.* **2014**, *30*, 117–120.
21. Padua, M.G. Designing an identity: The synthesis of a post-traditional landscape vocabulary in Hong Kong. *Landsc. Res.* **2007**, *32*, 225–240. [CrossRef]
22. Zhang, W.; Wang, Y.; Fu, L.; Hu, Y. Scenery Deconstruction: A New Approach to Understanding the Historical Characteristics of Nanjing Cultural Landscape. *Herit. Sci.* **2024**, *12*, 62. [CrossRef]
23. Jiang, Y.; Wang, S.; Li, X.; Hu, M. West Lake: A Model of Pursuit and Practice of an Ideal Habitat. *City Plan. Rev.* **2023**, *47*, 121–122.
24. Chen, Z. Place and Space: A Review of Landscape Anthropology Research. *Landsc. Archit. Front.* **2017**, *5*, 8–23.
25. Omohundro, J.T. *Thinking Like an Anthropologist: A Practical Introduction to Cultural Anthropology*, 1st ed.; McGraw-Hill: London, UK, 2007.
26. Bergeron, J.; Paquette, S.; Poullaouec-Gonidec, P. Uncovering landscape values and micro-geographies of meanings with the go-along method. *Landsc. Urban Plan.* **2014**, *122*, 108–121. [CrossRef]
27. Berkes, F. *Sacred Ecology: Traditional Ecological Knowledge and Resource Management*; Taylor & Francis: Oxfordshire, UK, 1999.
28. Bruno, D.M.; Kelly, S. Cambodia's Tonle Sap Lake: Iterating Between Disruptive Water Engineering and TEK. *Landsc. Archit.* **2022**, *29*, 12–25.
29. Awuah-Nyamekye, S. Indigenous Ways of Creating Environmental Awareness: A Case Study from Berekum Traditional Area, Ghana. *J. Study Relig. Nat. Cult.* **2014**, *8*, 46–63. [CrossRef]
30. Lam, D.P.M.; Hinz, E.; Lang, D.J.; Tengö, M.; von Wehrden, H.V.; Martín-López, B. Indigenous and local knowledge in sustainability transformations research: A literature review. *Ecol. Soc.* **2020**, *25*. [CrossRef]
31. Burchardt, J. From field walking to phenomenology: A review of recent British landscape historiography. *Hist. J.* **2024**, *67*, 583–609. [CrossRef]
32. Hoskins, W.G. *The Making of the English Landscape*; Little Toller Books: Southampton, UK, 2013.
33. Bodenhamer, D.J.; Beyond GLS: The Promise of Spatial Humanities[EB/0L]. West Lafayette: Presentation in Purdue University. 7 November 2014. Available online: https://docs.lib.purdue.edu/purduegisday/2014/Presentations/1/ (accessed on 24 October 2024).
34. Zhao, X.; Chen, S.; Wang, X.; Lin, Q. Perception of "City-Lake" system's spatial scope and landscape formation with regional orientation: Taking Huizhou West Lake as an example. *Chin. Landsc. Archit.* **2024**, *40*, 92–98.
35. Yang, C.; Sim, J.; Lawson, G. Deciphering Historic Landscapes: A Case Study of Slender West Lake in Yangzhou, China. *Landsc. Res.* **2016**, *41*, 95–112. [CrossRef]
36. Duan, S.; Lin, Q. Integration of city and landscape: Yinchuan traditional regional landscape system. *Chin. Landsc. Archit.* **2022**, *38*, 38–43.
37. Wang, Z.; Lin, Q.; Wang, X. Study on traditional landscape characteristics and landscape system along the eight Taihang passages. *Chin. Landsc. Archit.* **2024**, *40*, 28–34.

38. Jiang, X.; Lin, Q. Research on the features of traditional landscape system in the area along the Huaiyang canal under the influence of the canal water conservancy. *Chin. Landsc. Archit.* **2022**, *38*, 34–39.
39. Zhang, X. Historical Research of Xuchang West Lake, a famous scenic spot in Tang and Song Dynasty. *Song Hist. Res. Ser.* **2020**, *1*, 174–188.
40. Liang, S. Creating the sceneries by literature: Exploring the culture connotation of "8 sceneries" and "12 sceneries" of Hui Zhou West Lake. *Archit. Cult.* **2017**, *7*, 72–73.
41. Yan, M. From "Dongpo Model" to "West Lake Image": Su Shi's conception and practice of ecological agriculture. *Agric. Archaeol.* **2021**, *1*, 34–40.
42. Chen, W. Study on the construction by and landscaping of Su Causeway by Su Shi of Song Dynasty. *Landsc. Archit.* **2019**, *26*, 114–118.
43. Wang, J. Research on the water conservancy and culture landscape projects of Hangzhou West Lake presided by Su Dongpo. *Chin. Landsc. Archit.* **2018**, *34*, 14–18.
44. Fu, H. How to take the international road of Hangzhou tourism. *Obs. Ponder.* **2004**, *19*, 28–30.
45. Li, H. West Lake world heritage and Hangzhou tourism development practice and thinking. *Tour. Trib.* **2012**, *27*, 12–13.
46. Xu, S.; Liu, Y.; Qian, Y.; Wang, Q. The significance of the West Lake pattern and its heuristic implications for creating China's heritage tourism economics. *Tour. Manag.* **2017**, *58*, 286–292. [CrossRef]
47. Guo, W.; Sun, J.; Ji, W.; Hou, X. Landscape of the Irrigation Districts: Traditional Landscape Camping in the Perspective of Territorial Spatial Development. *J. Hum. Settl. West China* **2023**, *38*, 17–23.
48. Guo, W.; Wu, D.; Hou, X. Research on Traditional Irrigation Districts in China from the Perspective of Human Habitat Environment. *Chin. Landsc. Archit.* **2021**, *37*, 11–15.
49. Mai, H. Gardening time: Horticulture and meditations on time in Song (960–1279) China. *Postmedieval* **2023**, *14*, 749–771. [CrossRef]
50. Mao, H.; Wu, X. An Analysis on Developments of Urban Parks in Song Dynasty. *J. Hum. Settl. West China* **2013**, *28*, 11–16.
51. Yuan, P. Inheritance and Evolution of West Lake's Context in Modern Scenic Area—Take Yingzhou West Lake for Reference. Master's Thesis, Beijing Forestry University, Beijing, China, 2010.
52. Liao, R.; Hou, W. The government park in Tang Dynasty: Chongzhou Yanhuachi Garden. *Chin. Landsc. Archit.* **2004**, *10*, 14–21.
53. Chen, R.X. Zhangpu Prefectural Chronicle. 1968. Available online: http://read.nlc.cn/menhu/allSearch/searchDetail?searchType=all&showType=1&indexName=data_403&fid=312001076894 (accessed on 22 December 2024).
54. Zhang, X.; Wang, X. Participation of Man in Nature: Study on the Hill and Water Landscape System in the City of Fuzhou. *Landsc. Archit.* **2018**, *25*, 34–39.
55. Guo, W.; Hou, X. Two cities, three mountains and river net—Analysis of Fuzhou's landscape pattern and traditional urban form. *Landsc. Archit.* **2017**, *5*, 94–100.
56. Wang, X.; Wang, X. The landscape system and characteristics of ancient impounding lake from the "Water Conservancy-Landscape" perspective. *Landsc. Archit.* **2021**, *28*, 74–79.
57. Bennett, K.H.; Buck, C.E. Interpretation of lake sediment accumulation rates. *Holocene* **2016**, *26*, 1092–1102. [CrossRef]
58. Han, F. The Chinese View of Nature: Tourism in China's Scenic and Historic Interest Areas. Ph.D. Thesis, Queensland University of Technology, Brisbane, Australia, 2006.
59. Lin, Y. *My Country and My People*; Foreign Language Teaching and Research Press: Beijing, China, 2002.
60. Zong, Y. "Share the Pleasures with the Public": Evolution of Urban Landscapes in Nanjing in Song Dynasty China. *Cult. Hist.* **2024**, *21*, 1–22. [CrossRef]
61. Zhang, X. Study on the Hill and Water Landscape System in the City of Fuzhou. Ph.D. Thesis, Beijing Forestry University, Beijing, China, 2018.
62. Oberkircher, L.; Shanafield, M.; Ismailova, B.; Saito, L. Ecosystem and social construction: An interdisciplinary case study of the Shurkul lake landscape in Khorezm, Uzbekistan. *Ecol. Soc.* **2011**, *16*, 20. [CrossRef]
63. Li, M.; Zhou, Y.; Gong, D.; Dong, J.; Yan, H. Incorporation mechanism of traditional Chinese garden elements in modern urban landscape design: An exploratory analysis based on grounded theory. *J. Asian Archit. Build. Eng.* **2024**, 1–15. [CrossRef]
64. Yang, C.; Lawson, G.; Sim, J.; Han, F. Towards an understanding of dialectical authenticity of historic landscapes in China. *Geogr. Res.* **2020**, *59*, 285–300. [CrossRef]

land

MDPI

Article

The Influence of Multi-Sensory Perception on Public Activity in Urban Street Spaces: An Empirical Study Grounded in Landsenses Ecology

Tianqi Han [1,2], Lina Tang [3], Jiang Liu [1,2,*], Siyu Jiang [3,4] and Jinshan Yan [3,4]

1 School of Architecture and Urban-Rural Planning, Fuzhou University, Fuzhou 350108, China; 221520007@fzu.edu.cn
2 Fujian Key Laboratory of Digital Technology for Territorial Space Analysis and Simulation, Fuzhou 350108, China
3 Key Laboratory of Urban Environment and Health, Institute of Urban Environment, Chinese Academy of Sciences, Xiamen 361021, China; lntang@iue.ac.cn (L.T.); syjiang@iue.ac.cn (S.J.); jsyan@iue.ac.cn (J.Y.)
4 University of Chinese Academy of Sciences, Beijing 100049, China
* Correspondence: jiang.liu@fzu.edu.cn

Abstract: The design of street spaces significantly influences public behavior and quality of life. Understanding how various urban street spatial characteristics affect public behavior, alongside the role of multi-sensory perception, enables designers and planners to create more human-centered urban environments. Grounded in landsenses ecology, this study employs correlation analysis, regression analysis, and Partial Least-Squares Structural Equation Modeling (PLS-SEM) to examine the effects of different urban street spatial characteristics on public behavior and the mediating role of multi-sensory perception. The findings reveal that street spatial characteristics, particularly the Water Surface Ratio (WSR) and Waterfront Density (WD), have a pronounced impact on behavioral traits, with higher public activity frequencies in areas with elevated WSR and WD. Notably, WSR significantly affects static behaviors, such as sunbathing ($\beta = 0.371$, $p < 0.001$), and dynamic behaviors, such as walking ($\beta = 0.279$, $p < 0.001$). While road and water characteristics directly influence behavior, buildings and green spaces mainly affect public behavior through multi-sensory perception. Different sensory perceptions show varying effects, with olfactory perception playing a significant role in these experiences, alongside a notable chain-mediated effect between tactile perception and psychological cognition. These results provide valuable insights for integrating multi-sensory experiences into urban design.

Keywords: landsenses evaluation indices; street space characteristics; PLS-SEM; mediation effect

1. Introduction

Cities are confronted with urgent health, social, and environmental challenges, as emphasized by the United Nations Sustainable Development Goals (SDGs) [1]. Urban planning plays a critical role in shaping the spatial structure of cities, influencing lifestyles and environmental exposures that ultimately affect human health and sustainable development [2]. As the concept of "One Health" gains global recognition, the connection between sustainable urban environments and public health becomes increasingly significant. This shift necessitates a focus on human well-being, education, and sustainable living practices [3]. Streets, as vital public spaces within urban life, promote a positive perception that maintains a continuous relationship between pedestrians and their surroundings [4]. This relationship is essential for enhancing the quality of the living environment.

Urban streets are defined as roadways that facilitate daily social interactions among residents in towns and cities, along with the surrounding buildings, associated vegetation, and other facilities [5]. As fundamental components of the urban built environment, streets significantly influence individual behaviors through their transportation, social, and aesthetic functions [6]. Streets serve as some of the most frequented venues for walking, cycling, and various recreational activities [7]. Research indicates that high-quality street environments effectively promote active engagement among residents [8] and significantly influence public behavior characteristics [9], behavioral satisfaction [10], and route selection [11]. Street greenery significantly impacts both walking frequency and duration [9,12], while elements of the street environment, including pavement and facilities, influence pedestrians' walking preferences [13,14]. Although land use types significantly impact people's behavior, their influence is indirect, typically mediated by the physical characteristics of street space. For instance, wider roads typically incorporate more sidewalks and commercial facilities, which directly facilitate pedestrian activity [15]. In contrast, areas near water offer a peaceful environment, encouraging people to engage in activities such as walking [16]. Therefore, the physical characteristics of street space have a more direct influence on people's behavioral choices. However, most existing studies primarily focus on the relationship between street environments and walking behavior, leading to the insufficient exploration of other behavioral characteristics associated with streets. Previous studies have classified activities into static and dynamic behaviors based on varying levels of physical activity. Static behaviors are characterized by lower levels of physical exertion and energy expenditure, including actions such as resting, reading, conversing, or observing within a specific space [11]. In contrast, dynamic behaviors involve higher levels of physical activity, including spontaneous movements in specific areas, sports, and activities that rely on fitness and recreational facilities [17]. Understanding how the spatial characteristics of streets influence public behavior will help identify effective strategies for enhancing urban environments and improving residents' quality of life.

Various objective environmental characteristics offer the public a range of spatial experiences, triggering complex perceptions and behavior [17]. Scholars have employed the Stimulus–Organism–Response (SOR) model to clarify a series of behavioral responses elicited by the objective environment as a source of stimuli [18]. Landsenses ecology emphasizes individuals' perceptual experiences in urban environments, focusing on the systematic nature of multi-sensory perception and the holistic aspects of psychological cognition. This underscores the necessity of investigating land use planning, construction, and management from multiple related perspectives, including natural elements, physical perception, psychological cognition, socio-economics, processes, and risks [19]. Research indicates that multi-sensory perception is a fundamental condition for enhancing residents' satisfaction in urban environments, while psychological cognition serves as a direct factor influencing this satisfaction [20]. Furthermore, characteristics of visual and auditory perception significantly affect tourists' experiences [21], while sensory perceptions—including visual, auditory, and olfactory—have notable effects on psychological restoration [22,23]. However, existing studies have paid inadequate attention to the impacts of multi-sensory perception and psychological cognition on public activity characteristics. As significant mediators between the environment and behavior, perceptual information influences individuals' cognitive and emotional experiences of space through various channels, including vision, hearing, smell, and touch, thereby directly shaping their behavioral decisions and activity patterns [4]. This influence not only determines the duration of individuals' stays in specific environments, the types of activities they engage in, and their frequency of activity but also reflects the close relationship between psychological states and the environment. Understanding these mediating roles can aid in designing environments that more effectively meet individuals' sensory needs, thereby enhancing the overall user experience.

Urban street spaces serve as direct facilitators of individuals' perception and understanding of the urban environment. Comprehending the interplay between the components of urban streets and public behavior characteristics is essential for constructing health-centered cities that prioritize human well-being [24]. However, while the characteristics of street spaces, multi-sensory perception, and psychological cognition significantly influence public behavior, comprehensive assessments of their interactions and effects on behavior are still limited. Although some frameworks have identified pathways between physical environmental characteristics and behavioral traits [25,26], few studies have assessed the mediating roles of multi-sensory perception and psychological cognition. Furthermore, existing research often focuses on isolated sensory perceptions. Human sensory perception operates as an integrated system; thus, it is crucial to examine both individual sensory experiences and the cumulative effects of multi-sensory perception. Psychological cognition is fundamentally based on multi-sensory perception, as it synthesizes information from diverse sensory modalities, enabling individuals to form a comprehensive understanding of their environment [27]. This synthesis not only influences emotional responses and decision-making processes but also shapes behavioral outcomes [28,29]. Therefore, examining the mediating roles of multi-sensory perception and psychological cognition is crucial for gaining a deeper understanding of how street spatial characteristics affect public behavior.

This study focuses on street spatial characteristics across four categories: road, building, green space, and water. It quantifies relevant spatial data to explore how these features influence public behavioral traits, with multi-sensory perception and psychological cognition serving as mediators, as illustrated in Figure 1. This research addresses the following scientific questions: (1) What key spatial features influence people's behavioral traits on urban streets? (2) How could multi-sensory perception and psychological cognition shape the pathways through which urban street spatial characteristics impact people's behavior? This study could provide a theoretical foundation for improving urban space design and enhancing residents' quality of life.

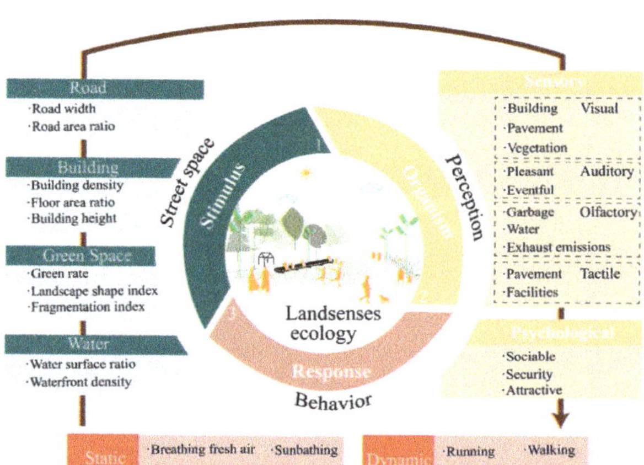

Figure 1. The influence of the spatial characteristics of urban streets on behavioral traits is examined from the perspective of landsenses ecology.

2. Method

2.1. Study Area and Sampling Points

The study was conducted in Fuzhou, a coastal city situated in the eastern part of Fujian Province, China, with a typical subtropical monsoon climate characterized by warm and humid conditions, as well as greenery present throughout the year. The research area is located in Gulou District, a central urban area in the northwest section of Fuzhou integrating various functional areas, such as historical districts, parks, residential areas, and commercial zones. This area features significant landmarks, including Wushan Mountain, the Three Lanes and Seven Alleys historical district, and Xihu Park. Since 2012, the research area has implemented a slow transportation system, with plans for a network that incorporates riverfront, waterfront, lakeside, and park connectivity, aimed at creating a distinctive "Min culture slow life" experience. Considering data availability and the feasibility of field research, this study selects 11 service-oriented streets and 8 landscape-oriented streets as subjects for analysis. The selection criteria for the sites are as follows: (1) The selection of sample sites considered various spatial environmental factors, including roads, buildings, green spaces, and water characteristics, ensuring that the spatial characteristics of the sites varied, thereby guaranteeing their representativeness. (2) To minimize the influence of any single sensory factor (such as traffic noise), the site selection process emphasized a balanced multisensory experience. Finally, based on the field survey, sites that were under construction or deemed unsuitable for research were excluded, resulting in a total of 40 viable research sites, as shown in Figure 2.

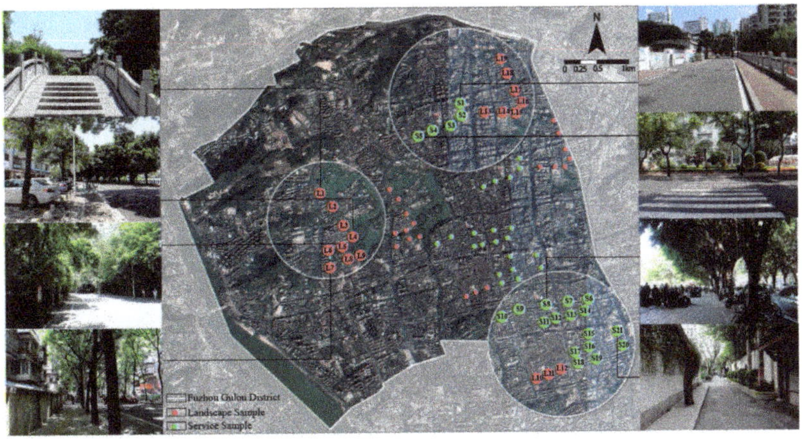

Figure 2. Spatial distribution map of the study area and sample points.

2.2. Data on Street Spatial Features

The research area was categorized into four types of spatial environmental elements—roads (including urban roads, internal roadways, sidewalks, and hard surfaces), buildings, green space, and rivers—using Google satellite imagery of the Gulou District from 2022 (Google Earth). Road network data for the Gulou District of Fuzhou City were obtained from Open Street Map while building outline data and the number of floors were acquired using Local Space Viewer software (version V4.2.0). To analyze the key spatial features influencing landscape perception evaluations of streets in the research area and to account for the scale of sensory perception, buffer zones with a radius of 100 m were established around selected points for survey and indicator calculations [30]. The methods for quantifying urban street space indicators are detailed in Table 1.

Table 1. Quantification of the urban street space indicators.

Elements	Indicators	Calculation Formula	Explanation
Road	Road Width (RW)	$RW = D_v + D_s$	D_v denotes the width of the vehicle lane. D_s represents the width of the sidewalk.
	Road Area Ratio (RAR)	$RAR = \frac{S_r}{S}$	S_r denotes the road surface area within the buffer zone. S denotes the total area within the buffer zone. RAR calculation in this study mainly includes urban main roads and side roads, excluding roads within residential areas.
Building	Building Density (BD)	$BD = \frac{S_g}{S}$	S_g denotes the ground floor area within the buffer zone.
	Floor Area Ratio (FAR)	$FAR = \frac{S_b}{S}$	S_b denotes the total building area within the buffer zone.
	Building Height (BH)	$BH = \frac{\sum_{i=1}^{n} H_i}{n}$	H_i denotes the height of the i-th building within the buffer zone, while n denotes the number of buildings within the buffer zone.
Green Space	Green Rate (GR)	$GR = \frac{S_g}{S}$	S_g denotes the total area of green space within the buffer zone.
	Landscape Shape Index (LSI)	$LSI = \frac{0.25C_g}{\sqrt{S_g}}$	C_g denotes the total perimeter of green space within the buffer zone.
	Fragmentation Index (FI)	$FI = \frac{N_i}{\sum S_i}$	N_i denotes the number of patches within the buffer zone corresponding to the i-th green area, while S_i represents the area of the i-th green space within the buffer zone.
Water	Water Surface Ratio (WSR)	$WSR = \frac{S_w}{S}$	S_w denotes the total area of water within the buffer zone. WSR refers to the proportion of the water surface area within the buffer zone—including lakes, rivers, and artificial ponds—relative to the total area of the buffer zone.
	Waterfront Density (WD)	$WD = \frac{C_w}{S}$	C_w denotes the perimeter of the water's edge within the buffer zone.

2.3. Public Survey Data

2.3.1. Landsenses Evaluation Indices of Urban Streets

Urban streets serve as essential environments for community activities, where environmental factors such as light, sound, smell, color, shape, and texture significantly influence individual perceptions and behaviors. This study establishes an evaluation index system aimed at assessing the perception of urban streets based on their spatial characteristics. Drawing on established landsenses indicators [29,31,32], the system encompasses five dimensions—visual, auditory, olfactory, tactile, and psychological—and integrates landsenses elements associated with roads, buildings, green spaces, and water, as shown in Table 2.

Each of these dimensions is designed to capture specific sensory experiences and their impact on individual psychology. Previous studies typically categorize the street interface into four key components: canopy, building wall, ground plane, and roadside plane [33]. This study primarily focuses on visual perception, particularly factors such as building design, paving material color, and vegetation richness, which collectively shape the street's visual impression and influence pedestrians' emotional responses and behavioral choices. Auditory perception, on the other hand, emphasizes the evaluation of the pleasantness and eventfulness of the street's soundscape [34], particularly focusing on how these auditory characteristics influence pedestrian behavior patterns. Olfactory perception primarily concerns odors. While positive odors can enhance the overall environmental perception [35], urban streets are frequently exposed to negative smells such as trash, waterborne odors, and vehicle exhaust. This study specifically addresses how these factors significantly influence the perception of street quality. Tactile perception focuses on the physical characteristics of the street, particularly the smoothness of the pavement, the comfort of resting facilities, and their ability to meet pedestrian needs [36]. Street design should focus on optimizing these sensory experiences, providing an environment that is not only safe and comfortable but also attractive, thereby promoting positive pedestrian behavior [33]. In conclusion, the aim of this study is to comprehensively understand how the sensory elements of urban streets shape public behavior and influence the overall perception of space, thereby providing a theoretical basis for optimizing street design.

Table 2. Landsenses indicators of urban street.

Sensory Perception	Characterization Indicators
Visual	The building is colorful. The pavement color is very comfortable. The vegetation is very lush.
Sound	The soundscape is pleasant. The soundscape is eventful.
Olfactory	Odor of the garbage. Odor of the water. Odor of automobile exhaust.
Tactile	The pavement is smooth. The rest facilities are well-equipped. The rest facilities are comfortable.
Psychological	The atmosphere of interaction is well. It has a sense of security. It is attractive.

2.3.2. Questionnaire Design

The public survey data for this study were collected using a structured questionnaire. The questionnaire consisted of three main sections. The first section gathered basic demographic information about the respondents. The second section focused on the evaluation of sensory experience and psychological perception related to the spatial characteristics of the streets. Drawing on the previously established evaluation index system for landsenses ecology, respondents were instructed to assess their environment after a one-minute observation across five dimensions: visual, auditory, olfactory, tactile, and psychological, using a scale ranging from 1 (strongly disagree) to 7 (strongly agree). The third section focused on the types and frequencies of the respondents' activities. Based on observations of the behaviors of street users, these activities were classified as static or dynamic, as shown in Table 3. Respondents were asked to indicate the frequency of their activities using a seven-point Likert scale, ranging from 1 (rarely) to 7 (very frequently). The survey utilized a random sampling method. Prior to participation, respondents were thoroughly informed of the study's objectives and gave their consent. Participants were subsequently guided through the completion of the questionnaire. To ensure the representativeness of the sample, respondents were selected from a broad spectrum of demographic backgrounds, including gender, age, and occupation, thereby improving the generalizability and applicability of the results. The random selection and demographic diversity of the sample bolstered the reliability and validity of the study's findings. To account for crowd activity and ensure the feasibility of the survey, data collection was conducted on clear weather weekdays during three specific time periods: from 8:00 to 11:00 a.m., from 2:00 to 5:00 p.m., and from 5:00 to 8:00 p.m. Ultimately, a total of 471 valid questionnaires were collected from 40 sampling sites.

Table 3. Classification of behavior types in urban streets.

Behavior Type	Concrete Behavior
Static activities	breathe the fresh air, sunbathe, chat, play cards.
Dynamic activities	run, walk, walk the dog, shopping

2.4. Construction of the Conceptual Model

This study investigates four categories of street spatial characteristics: buildings, roads, green spaces, and water. From the perspective of landsenses ecology, this research aims to quantify individuals' multi-sensory perception and psychological cognition while investigating the mechanisms through which street spatial characteristics influence behavior. The model is designed to illustrate the pathways through which the four categories of street spatial characteristics impact the two dimensions of activity types, with street spatial characteristics serving as independent variables, behavioral traits as dependent variables, and multi-sensory perception and psychological cognition as mediators. A structural equation model is proposed, as illustrated in Figure 3, along with the following specific research hypotheses:

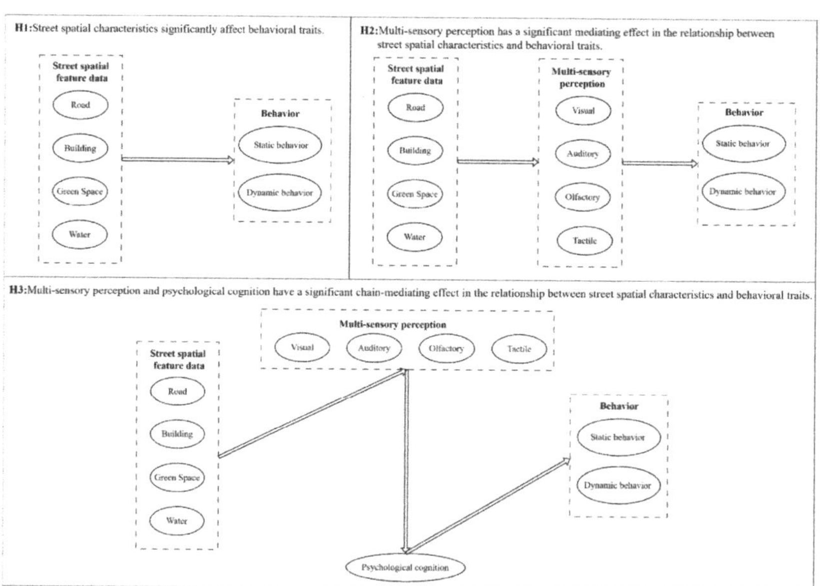

Figure 3. Impact pathways of urban street spatial characteristics on behavioral outcomes.

H1: Street spatial characteristics significantly affect behavioral traits.

H2: Multi-sensory perception has a significant mediating effect in the relationship between street spatial characteristics and behavioral traits.

H3: Multi-sensory perception and psychological cognition have a significant chain-mediating effect in the relationship between street spatial characteristics and behavioral traits [26,29].

This study employed Partial Least Squares-Structural Equation Modeling (PLS-SEM) as the statistical technique. PLS-SEM is utilized to test the hypotheses, offering distinct advantages over traditional structural equation models, particularly for smaller sample sizes and exploratory research [37]. This multivariate analysis comprises two interconnected components: the measurement model, which assesses reliability and validity to confirm the predictive potential of observed variables for their latent constructs, and the structural model, which evaluates the predictive relationships among these constructs through hypothesis testing and mediation analysis. The sample size for this study ($N = 471$) exceeds the recommended minimum of ten times the number of questionnaire items (32 items), thereby ensuring reliable outcomes from the PLS-SEM analysis. Although PLS-SEM provides substantial advantages, several limitations need to be addressed. Firstly, even though the sample size surpassed the recommended threshold, its representativeness might still restrict the broader applicability of the findings [37]. Moreover, PLS-SEM necessitates a balance between sample size adequacy and the robustness of standardized model evaluation criteria, which highlights areas requiring methodological refinement [38]. Statistical analysis was carried out using Smart PLS 3.0.

3. Results

3.1. Basic Characteristics of Urban Street Spatial Features

The spatial characteristics of the streets in the study area are detailed in Table 4. The average Road Width (RW) is 13.9 m, with the majority of streets designed with two lanes to accommodate bidirectional traffic. The average Building Height (BH) is approximately 24 m, primarily comprising seven-story residential structures, and the average Building Density (BD) is 28.5%, reflecting an appropriate level of land use intensity. The average Green Rate (GR) is 35.5%, and the low fragmentation value suggests that the green spaces in the study area are relatively intact, indicating a high level of green space development. Few roads are located near rivers, and both the Water Surface Ratio (WSR) and Waterfront Density (WD) are low.

Table 4. Statistical analysis of street space features.

Elements	Indicators	Mean	Max	Min	Standard Deviation
Road	RW/m	13.9	20.0	8.3	3.0
	RAR/%	41.9	66.6	12.5	13.4
Building	BD/%	28.5	45.4	8.6	8.7
	FAR	2.017	4.843	0.196	1.075
	BH/m	24.499	50.267	5.279	9.542
Green Space	GR/%	35.5	64.2	20.5	12.0
	LSI	11.367	20.732	5.417	3.713
	FI/%	0.106	0.299	0.009	0.074
Water	WSR/%	0.079	0.345	0.000	0.113
	WD/%	0.856	3.642	0.000	1.121

3.2. Sample Statistics

As shown in Figure 4, among the 471 valid responses, 267 (57%) are male, while 204 (43%) are female, yielding a gender ratio of approximately 1:1.3. The largest proportion of respondents falls within the age group of 25 to 30 years, comprising 21% of the total. Furthermore, 77% of respondents indicate that their commuting time to the survey location does not exceed 0.5 h.

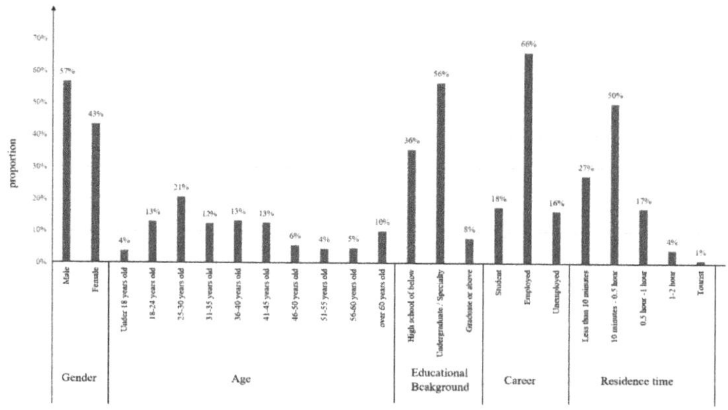

Figure 4. Composition analysis of the respondent group.

Individual characteristics (such as gender, age, etc.) are potential factors contributing to differences in the evaluation of street perception indicators. Given the non-normal distribution of the data, non-parametric tests are employed to investigate the difference in perception indicators among groups with varying individual characteristics and to identify statistically significant variations. The results, as summarized in Table 5, show that gender exhibits significant differences in Well-equipped facilities, Interaction, and Security, with males assigning higher scores than females. No significant differences are observed across age groups for any perception indicators. Educational background demonstrates significant differences in the evaluation of Sound Pleasantness, Garbage Odors, and Automobile Exhaust Odors, indicating variations in how individuals with different educational backgrounds evaluated these aspects. Similarly, people with different occupations show different opinions on Garbage Odors, Well-equipped Facilities, and Interaction, with unemployed individuals assigning lower scores to Garbage Odors but higher scores to Well-equipped Facilities and Interaction compared to students and employed individuals. Finally, residential distance shows significant differences in the perception of "the pavement color is very comfortable", with non-local visitors and those living within 30 min of the sampled street providing the highest ratings.

Individual characteristics also show significant variation in their association with public behavioral patterns in street environments. As presented in Table 6, gender shows significant differences in static behaviors, including chatting, playing cards, and sunbathing, with males engaging in these activities more frequently than females. Age shows notable differences across all behaviors except chatting and shopping, with individuals aged 51–60 being the most active group across most activities. Educational background exhibits significant differences across all behaviors. Respondents with undergraduate or specialty degrees are the most active ones in shopping, while those with higher education levels participate more frequently in other activities. Conversely, residential distance from the sampled site does not reveal any significant difference in behavioral patterns.

Table 5. Summary of the non-parametric test results examining the relationships between individual characteristics and landsenses indicators.

Indicator		Gender	Age	Educational Background	Career	Residence Time
Visual	Building	0.054	0.086	0.312	0.262	0.378
	Pavement	0.711	0.212	0.847	0.243	0.004 **
	Vegetation	0.194	0.576	0.247	0.192	0.851
Sound	Pleasantness	0.358	0.268	0.027 *	0.052	0.488
	Eventfulness	0.558	0.990	0.146	0.842	0.476
Olfactory	Garbage	0.590	0.094	0.024 *	0.012 *	0.878
	Water	0.632	0.316	0.381	0.936	0.521
	Automobile	0.64	0.263	0.042 *	0.871	0.360
Tactile	Pavement	0.551	0.101	0.058	0.171	0.177
	Well-equipped facilities	0.039 *	0.091	0.063	0.033 *	0.878
	Comfortable facilities	0.268	0.274	0.186	0.440	0.787
Psychological	Interaction	0.045 *	0.125	0.318	0.050 *	0.748
	Security	0.007 **	0.349	0.207	0.488	0.712
	Attraction	0.257	0.247	0.187	0.076	0.279

* $p < 0.05$, ** $p < 0.01$, *** $p < 0.001$.

Table 6. Summary of non-parametric test results analyzing the relationships between individual characteristics and behavioral frequencies.

Behavior		Gender	Age	Educational Background	Career	Residence Time
Static activity	breath the fresh air	0.998	0.004 **	0.001 **	0.000 ***	0.287
	chat	0.024 *	0.138	0.001 **	0.575	0.840
	play cards	0.026 *	0.002 **	0.000 ***	0.000 ***	0.487
	sunbathe	0.008 **	0.011 *	0.000 ***	0.005 **	0.460
Dynamic activities	walk the dog	0.952	0.006 **	0.005 **	0.015 *	0.501
	run	0.282	0.015 *	0.012 *	0.092	0.348
	walk	0.829	0.002 **	0.012 *	0.000 ***	0.552
	shopping	0.441	0.061	0.032 *	0.024 *	0.668

* $p < 0.05$, ** $p < 0.01$, *** $p < 0.001$.

3.3. Relationship Between Street Spatial Characteristics and Behavioral Traits

As shown in Figure 5, regarding road characteristics, Road Width (RW) is positively correlated with activities such as enjoying fresh air, sunbathing, and walking but negatively correlated with shopping. The Road Area Ratio (RAR) is negatively correlated exclusively with running, while it is positively correlated with shopping. In terms of building characteristics, Building Density (BD) does not show a significant correlation with chatting; however, it is significantly negatively correlated with other behaviors, except for shopping. The correlation results for the Floor Area Ratio (FAR) align with those of Building Density (BD), indicating no significant correlation with either chatting or sunbathing. The average Building Height (BH) is significantly positively correlated only with chatting and sunbathing. Concerning green space, the Green Rate (GR) is positively correlated with running but shows a significant negative correlation with both chatting and shopping. The Landscape Shape Index (LSI) demonstrates a positive correlation with chatting, walking, and shopping behaviors. The Fragmentation Index (FI) is negatively correlated with activities including enjoying fresh air, sunbathing, walking dogs, running, and walking, while it shows a positive correlation with shopping. Regarding water, both the Water Surface

Ratio (WSR) and Waterfront Density (WD) are significantly positively correlated with all behaviors except shopping, which shows a negative correlation.

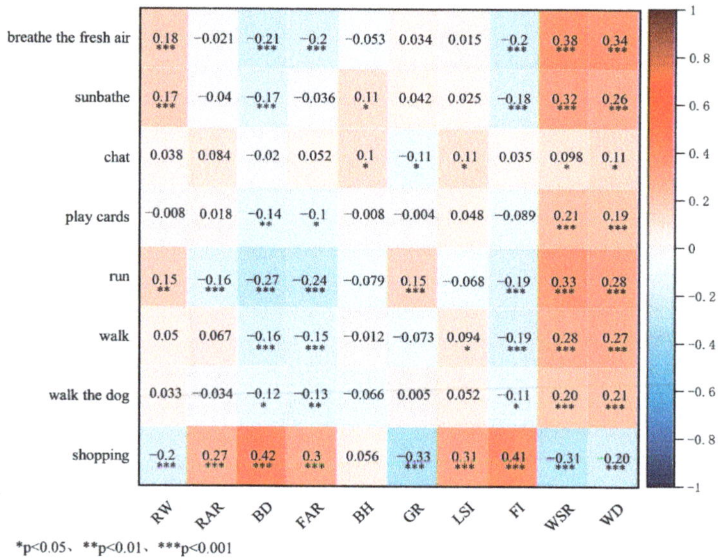

Figure 5. Correlation analysis results between the spatial characteristics of urban streets and behavioral traits.

3.4. Influence of Street Spatial Characteristics on Behavior Traits

3.4.1. Factors Influencing the Impact of Street Spatial Characteristics on Behavioral Traits

To further investigate the influence of street spatial characteristics on behavior, a multiple regression analysis is conducted, treating behavioral traits as the dependent variable and street spatial characteristics as the independent variable, as shown in Table 7. The results indicate that water characteristics could significantly influence the frequency of public behavior. Road characteristics have a limited effect, influencing only the activities associated with enjoying fresh air and shopping. Building characteristics are related to activities such as enjoying fresh air, running, walking, and shopping. Green space could significantly affect the frequency of all behaviors, except for dog walking and running.

Table 7. Multiple stepwise regression analysis of street spatial characteristics and behavior.

Type	Dependent Variable	Independent Variable	Beta	VIF	t	R^2	F
Static activity	Breathe the fresh air	RW	0.138	1.479	2.716 **	0.186	21.226 ***
		FAR	−0.137	1.68	−2.525 *		
		GR	−0.186	1.41	−3.738 ***		
		WSR	0.213	3.134	2.884 **		
		WD	0.147	2.508	2.214 *		
	chat	GR	−0.134	1.005	−2.944 ***	0.039	9.397 ***
		WD	0.154	1.005	3.387 **		
	play cards	WD	0.268	1.000	6.020 ***	0.072	36.242 ***
	sunbathe	GR	−0.116	1.073	−2.588 **	0.129	34.533 ***
		WSR	0.371	1.073	8.299 ***		

Table 7. *Cont.*

Type	Dependent Variable	Independent Variable	*Beta*	*VIF*	*t*	R^2	*F*
Dynamic activities	walk the dog	WD	0.229	1.000	5.102 **	0.053	26.028 ***
	Run	BD	−0.206	1.159	−4.388 ***	0.108	2.128 ***
		WD	0.191	1.159	4.059 ***		
	walk	FAR	−0.194	1.415	−3.823 ***	0.150	27.500 ***
		GR	−0.284	1.261	−5.92 ***		
		WSR	0.279	1.215	5.939 ***		
	shopping	RAR	0.149	1.261	3.239 **	0.218	32.386 ***
		BD	0.198	2.287	3.2 **		
		GR	−0.129	1.497	−2.58 *		
		WSR	−0.14	1.793	−2.556 *		

* $p < 0.05$, ** $p < 0.01$, *** $p < 0.001$.

3.4.2. The Impact Pathways of Street Spatial Characteristics on Behavioral Traits

When evaluating the Structural Equation Model (SEM), the quality of the model is primarily assessed through reliability and validity tests. The evaluation typically involves Composite Reliability (CR) and Average Variance Extracted (AVE) for each latent variable to gauge the model's reliability and validity [37]. After normalizing the data, we conducted reliability and validity analyses and found that the CR coefficients exceeded 0.8, and the AVE values were all above 0.6. These results indicate that the model demonstrates good internal consistency and convergent validity.

To meet the requirements for structural validity, the factor loadings of all indicators (representing the influence of observable variables on latent variables) must exceed 0.55 [37,39]. However, the factor loadings for the RAR, BH, LSI, and FI are below this threshold. To enhance the model's validity, we consider excluding these indicators. Additionally, the indicators for walking and shopping do not meet the established criteria and are subsequently removed from the model.

In terms of direct effects, only road and water characteristics significantly influence both static and dynamic behaviors. As shown in Table 8, the 95% confidence intervals for the estimated indirect effects do not include zero, indicating that there is a mediating effect of multi-sensory perception within this relationship.

Table 8. Results of the mediation effects of multi-sensory perception on the relationship between street spatial characteristics and behavioral traits.

Dependent Variable	Independent Variable	Direct Effect			Indirect Effect		
		Estimates (β)	*p*	95%CI	Estimates (β)	*p*	95%CI
Static activity	Road	−0.091	0.023	[−0.172,−0.015]	0.141	0.001	[0.098,0.192]
	Building	0.032	0.652	[−0.108,0.169]	−0.087	0.008	[−0.155,−0.024]
	Green space	−0.075	0.080	[−0.158,0.010]	−0.095	0.001	[−0.151,−0.044]
	Water	0.192	0.001	[0.081,0.304]	0.176	0.001	[0.109,0.248]
Dynamic activities	Road	−0.100	0.022	[−0.187,−0.013]	0.097	0.000	[0.057,0.144]
	Building	−0.053	0.470	[−0.199,0.090]	−0.054	0.040	[−0.103,−0.001]
	Green space	0.020	0.716	[−0.083,0.129]	−0.055	0.015	[−0.101,−0.011]
	Water	0.127	0.039	[0.004,0.246]	0.121	0.001	[0.058,0.198]

Figure 6 illustrates the significant pathways of indirect effects. Roads exert a direct influence on both static and dynamic behaviors, alongside a positive indirect effect mediated by auditory, olfactory, and tactile perceptions. Although buildings do not exhibit a significant direct effect on static and dynamic behaviors, they negatively influence these behaviors indirectly through tactile perception. Green spaces do not significantly impact static and dynamic behaviors; however, they negatively affect static behaviors through auditory and tactile perceptions and have a negative influence on dynamic behaviors via tactile perception. Water has a direct effect on both static and dynamic behaviors, providing a positive indirect influence on static behaviors through auditory, olfactory, and tactile perceptions, as well as a positive indirect effect on dynamic behaviors through olfactory and tactile perceptions.

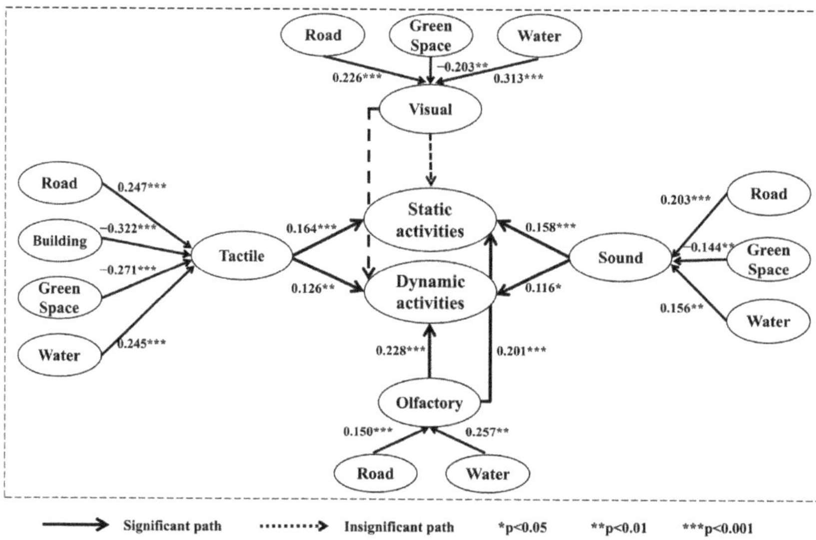

Figure 6. Influence pathways of urban street spatial characteristics on behavioral traits.

As shown in Table 9, further investigation into the mediating roles of multi-sensory perception and psychological cognition demonstrates that street spatial characteristics exert a chain-mediated effect on static behaviors through tactile perception and psychological cognition. Additionally, water influence psychological cognition through visual perception, which subsequently affects static behaviors.

Table 9. Significant pathways of chain mediation effects.

Influence Path	Estimates (β)	*p*	95%CI
Road -> tactile perception -> psychological cognition -> static activity	0.014	0.015	[0.004,0.027]
Building -> tactile perception -> psychosocial cognition -> static activity	−0.019	0.015	[−0.035,0.005]
Green space -> tactile perception -> psychological cognition -> static activity	−0.016	0.018	[−0.031,−0.005]
Water -> tactile perception -> psychological cognition -> static activity	0.014	0.025	[0.004,0.028]
Water -> visual perception -> psychological cognition -> static activity	0.006	0.045	[0.001,0.013]

4. Discussion

4.1. The Relationship Between Street Spatial Characteristics and Behavioral Traits

Street spatial characteristics significantly influence public behavioral traits, with the presence of water exerting a particularly pronounced effect on activity levels. Environments featuring water are generally more appealing to the public [40], aligning with prior research that indicates spaces near water are more attractive and that water plays a significant role in influencing static behaviors in parks [16,17]. Conversely, the proportion of green space has a notable negative impact on public behavior. Previous studies have shown that the influence of urban green spaces is nonlinear; for instance, the visibility of green streetscapes and NDVI levels exhibit nonlinear effects on walking tendencies [41,42]. Moreover, the quantity of green spaces has a significant inverted U-shaped effect on life satisfaction, with an optimal proportion of 11% maximizing public life satisfaction [43]. In our case, the Green Rate (GR) ranges from 20.5% to 64.2%, which falls within the declining segment of this threshold. Therefore, while increasing green space may continue to positively influence behavior, exceeding the optimal ratio could diminish these beneficial effects and potentially result in negative outcomes. The adverse impact of the Green Rate (GR) is more pronounced on static behaviors, consistent with previous findings that visitors prefer lower per capita green space when seated compared to when walking [44].

According to the regression analysis results, both Road Width (RW) and Road Area Ratio (RAR) significantly and positively influence activity frequency. This supports earlier research indicating that Road Width (RW) is a critical variable influencing street activity [45], significantly impacting the fulfillment of walking needs [46], and showing a strong positive correlation with social interactions among middle-aged and elderly individuals [47]. Apart from shopping, building characteristics correlate negatively with behavior. Previous studies suggest a nonlinear relationship between Building Density (BD) and pedestrian traffic; specifically, in urban centers where Building Density (BD) exceeds 0.3, building characteristics negatively impact walking [48]. Future research should concentrate on the specific effects of spatial characteristics on behavior in high-density environments to enhance public activity experiences.

4.2. Pathways Through Which Street Spatial Characteristics Influence Behavior

Multi-sensory perception plays a crucial mediating role in the relationship between street spatial characteristics and behavioral traits. Specifically, roads and water not only directly influence both static and dynamic behaviors but also exert an indirect effect through multi-sensory perception. Conversely, building and green space characteristics primarily impact static and dynamic behaviors via multi-sensory perception. Previous research indicates that soundscapes significantly affect pedestrians' walking routes and behavioral patterns [49], influencing walking speed and extending dwell time [50,51]. Similarly, odors can affect crowd movement speed and dwell time [51]. Furthermore, pavement evenness, as well as the number and placement of resting facilities, are critical factors influencing outdoor activities [52]. These findings collectively underscore the significant influence of multi-sensory perception on behavior.

However, existing studies have primarily focused on the role of multi-dimensional physical perception in psychological restoration, with limited exploration of its effects on behavior. Prior research has highlighted the significant impact of visual, auditory, and olfactory senses on public restoration [22,23,53]. Among these, olfactory perception is particularly vital in the restoration process [53], supporting this study's conclusion that olfactory perception has the most substantial impact on behavioral activities. Humans possess a remarkable sense of smell, often surpassing that of many animals [54]. Research has demonstrated that in urban forest parks, children's olfactory experiences have the most significant influence on their behavior [55]. Moreover, when examining the effects of multi-sensory stimulation in urban green spaces on physiological stress, olfactory stimuli have been found to be more effective than visual stimuli in alleviating stress [56].

In the pathway of "street spatial characteristics—sensory perception—psychological cognition—behavioral traits", roads, buildings, green spaces, and water influence psychological cognition through tactile perception, which subsequently affects static behaviors. Tactile perception and psychological cognition serve as chain mediators in this process. Evidence indicates that multi-sensory stimulation from pleasant landscape features can lead to positive cognitive evaluations [28,29], thereby enhancing public behavior. Research from the perspective of landsenses ecology emphasizes the significance of tactile perception indicators [32,57]. For hearing-impaired visitors, tactile experiences are notably richer than those from other non-visual senses [58]. Furthermore, studies on the restorative effects of multi-sensory perception in urban green spaces suggest that tactile perception has an indirect relationship with psychological restoration, influencing residents' behavior and emotional responses, which in turn affects restoration outcomes [59]. This implies a closer connection between tactile perception and emotional cognition compared to other senses. In this study, individuals develop positive tactile perceptions through interactions with various paving materials and the use of resting facilities, fostering a strong cognitive atmosphere that influences behaviors such as playing cards and chatting. Thus, tactile perception plays a crucial role in understanding the environment and interacting with surrounding elements.

4.3. Research Limitations and Future Studies

While this study primarily focuses on the impact of street spatial characteristics on public behavior, land use and social activities may indirectly influence these behaviors by shaping spatial functions and patterns of human mobility. Future research could examine how these factors intersect with sensory experiences to shape public activity patterns in urban streets. Moreover, this study does not fully address the potential impacts of seasonal climate variations on pedestrian behavior and street space usage. Variations in temperature and humidity, in particular, may affect pedestrian comfort and, consequently, alter the types and frequency of activities in urban streets. Future studies could investigate how these climatic variations interact with human behavior and spatial usage, offering deeper insights into how climate shapes public engagement with urban environments.

5. Conclusions

Urban streets serve as essential public spaces that connect individuals to their environment, with spatial characteristics playing a pivotal role in shaping public perception and behavior. This study explores the influence of street spatial characteristics on public behavioral traits, emphasizing the mediating role of multi-sensory perception. Grounded in landsenses ecology, the research employs methodologies such as spatial analysis, correlation analysis, regression analysis, and Structural Equation Modeling (SEM). The key findings are summarized as follows:

(1) Among street spatial characteristics, water elements have the most substantial impact on public behavioral traits. Areas with higher Water Surface Ratios (WSRs) promote nature-based activities, while higher Waterfront Density (WD) fosters social and physical activities, such as chatting, running, and walking dogs.

(2) Roads and water features influence behaviors directly and indirectly through multi-sensory perception, particularly auditory, olfactory, and tactile modalities. Conversely,

buildings and green spaces affect behaviors primarily through sensory pathways, with certain characteristics, such as Building Density (BD) and Green Rate (GR), having negative impacts.

(3) Olfactory perception surpasses other sensory modalities in influencing public behavior. Unpleasant odors significantly deter public engagement, even in visually and acoustically appealing areas, highlighting the importance of managing odor sources in urban environments.

This study underscores the need to integrate multi-sensory perception into urban planning to enhance public engagement and well-being. Special attention should be given to olfactory landscapes and the interplay between tactile perception and psychological cognition. By optimizing these sensory dimensions, urban spaces can become safer, more attractive, and better connected to the communities they serve.

Author Contributions: Conceptualization, T.H., L.T. and J.L.; methodology, T.H. and J.L.; software, T.H.; validation, T.H., S.J. and J.Y.; formal analysis, T.H.; investigation, T.H. and S.Y; resources, J.L.; data curation, T.H.; writing—original draft preparation, T.H.; writing—review and editing, L.T., J.L., S.J. and J.Y.; visualization, T.H.; supervision, L.T. and J.L; project administration, J.L.; funding acquisition, J.L. All authors have read and agreed to the published version of the manuscript.

Funding: This work was supported by the National Natural Science Foundation of China (NO.52378049).

Data Availability Statement: Due to the privacy and confidentiality of the respondents, the questionnaire data used in this study cannot be made publicly available. However, the data can be accessed upon reasonable request and under the condition that participant privacy is ensured, by contacting the corresponding author.

Acknowledgments: We express gratitude to two individuals for their contributions: Master's student Mengqiao Zhang and Shiying Li from Fuzhou University provided assistance in data collection and article revision.

Conflicts of Interest: The authors declare no conflicts of interest.

References

1. UN General Assembly. *Resolution Adopted by the General Assembly: Transforming Our World: The 2030 Agenda for Sustainable Development A/RES/70/1*; United Nations: New York, NY, USA, 2015.
2. Lowe, M.; Adlakha, D.; Sallis, J.F.; Salvo, D.; Cerin, E.; Moudon, A.V.; Higgs, C.; Hinckson, E.; Arundel, J.; Boeing, G.; et al. City planning policies to support health and sustainability: An international comparison of policy indicators for 25 cities. *Lancet Glob. Health* **2022**, *10*, e882–e894. [CrossRef] [PubMed]
3. Ye, C.; Schröder, P.; Yang, D.; Chen, M.; Cui, C.; Zhuang, L. Toward healthy and liveable cities: A new framework linking public health to urbanization. *Environ. Res. Lett.* **2022**, *17*, 064035. [CrossRef]
4. Rezvanipour, S.; Hassan, N.; Ghaffarianhoseini, A.; Danaee, M. Why does the perception of street matter? A dimensional analysis of multisensory social and physical attributes shaping the perception of streets. *Arch. Sci. Rev.* **2021**, *64*, 359–373. [CrossRef]
5. Abusaada, H.; Elshater, A. Effect of people on placemaking and affective atmospheres in city streets. *Ain Shams Eng. J.* **2021**, *12*, 3389–3403. [CrossRef]
6. Tang, J.; Long, Y. Measuring visual quality of street space and its temporal variation: Methodology and its application in the Hutong area in Beijing. *Landsc. Urban Plan.* **2019**, *191*, 103436. [CrossRef]
7. Lu, Y. Using Google Street View to investigate the association between street greenery and physical activity. *Landsc. Urban Plan.* **2019**, *191*, 103435. [CrossRef]
8. Salazar-Miranda, A.; Heine, C.; Duarte, F.; Schechtner, K.; Ratti, C. Measuring the impact of slow zones on street life using social media. *Cities* **2022**, *131*, 104010. [CrossRef]
9. Lu, Y.; Sarkar, C.; Xiao, Y. The effect of street-level greenery on walking behavior: Evidence from Hong Kong. *Soc. Sci. Med.* **2018**, *208*, 41–49. [CrossRef]

10. Song, J.; Zhou, S.; Kwan, M.-P.; Liang, S.; Lu, J.; Jing, F.; Wang, L. The effect of eye-level street greenness exposure on walking satisfaction: The mediating role of noise and PM2.5. *Urban For. Urban Green.* **2022**, *77*, 127752. [CrossRef]
11. Liu, J.; Wei, Y.; Lu, S.; Wang, R.; Chen, L.; Xu, F. The elderly's preference for the outdoor environment in Fragrant Hills Nursing Home, Beijing: Interpreting the visual-behavioural relationship. *Urban For. Urban Green.* **2021**, *64*, 127242. [CrossRef]
12. Ki, D.; Lee, S. Analyzing the effects of Green View Index of neighborhood streets on walking time using Google Street View and deep learning. *Landsc. Urban Plan.* **2021**, *205*, 103920. [CrossRef]
13. Jin, L.; Lu, W.; Sun, P. Preference for Street Environment Based on Route Choice Behavior While Walking. *Front. Public Health* **2022**, *10*, 880251. [CrossRef]
14. Yan, Q.; Luo, S.; Jiang, J. Urban Residents' Preferred Walking Street Setting and Environmental Factors: The Case of Chengdu City. *Buildings* **2023**, *13*, 1199. [CrossRef]
15. Elzeni, M.M.; Elmokadem, A.A.; Badawy, N.M. Impact of urban morphology on pedestrians: A review of urban approaches. *Cities* **2022**, *129*, 103840. [CrossRef]
16. Yao, W.; Yun, J.; Zhang, Y.; Meng, T.; Mu, Z. Usage behavior and health benefit perception of youth in urban parks: A case study from Qingdao, China. *Front. Public Health* **2022**, *10*, 923671. [CrossRef] [PubMed]
17. Li, J.; Huang, Z.; Zhu, Z.; Ding, G. Coexistence Perspectives: Exploring the impact of landscape features on aesthetic and recreational values in urban parks. *Ecol. Indic.* **2024**, *162*, 112043. [CrossRef]
18. Mehrabian, A.; Russell, J.A. *An Approach to Environment Psychology*; MIT: Cambridge, MA, USA, 1974.
19. Zhao, J.; Liu, X.; Dong, R.; Shao, G. Landsenses ecology and ecological planning toward sustainable development. *Int. J. Sustain. Dev. World Ecol.* **2016**, *23*, 293–297. [CrossRef]
20. He, S.; Chen, D.; Shang, X.; Han, L.; Shi, L. Resident Satisfaction of Urban Green Spaces through the Lens of Landsenses Ecology. *Int. J. Environ. Res. Public Health* **2022**, *19*, 15242. [CrossRef] [PubMed]
21. Guo, X.; Liu, J.; Chen, Z.; Hong, X.-C. Harmonious Degree of Sound Sources Influencing Visiting Experience in Kulangsu Scenic Area, China. *Forests* **2023**, *14*, 138. [CrossRef]
22. Wei, Y.; Hou, Y. Forest Visitors' Multisensory Perception and Restoration Effects: A Study of China's National Forest Parks by Introducing Generative Large Language Model. *Forests* **2023**, *14*, 2412. [CrossRef]
23. Li, S.; Chen, T.; Chen, F.; Mi, F. How Does the Urban Forest Environment Affect the Psychological Restoration of Residents? A Natural Experiment in Environmental Perception from Beijing. *Forests* **2023**, *14*, 1986. [CrossRef]
24. Mehta, V.; Bosson, J.K. Revisiting Lively Streets: Social Interactions in Public Space. *J. Plan. Educ. Res.* **2018**, *41*, 160–172. [CrossRef]
25. Gao, W.; Jia, M.; Zhao, M.; Gao, Y.; Meng, H. Review of Progress and Quantitative Measurement Methods of Research on Street Space. *City Plan. Rev.* **2022**, *46*, 106–114.
26. Wang, Y.; Zhou, Q.; Yang, X.; Nan, J. Research on Process Framework and Evaluation System of Urban Public Spatial Perception. *Urban Plan. Int.* **2022**, *37*, 80–89.
27. Zhao, J.; Yan, Y.; Deng, H.; Liu, G.; Dai, L.; Tang, L.; Shi, L.; Shao, G. Remarks about landsenses ecology and ecosystem services. *Int. J. Sustain. Dev. World Ecol.* **2020**, *27*, 196–201. [CrossRef]
28. Liu, C.; Tang, L.; Yan, J.; Ouyang, J. Direct and indirect effects of multisensory modalities on visitor's thermal comfort in an urban park in a humid-hot climate. *Int. J. Sustain. Dev. World Ecol.* **2023**, *30*, 319–328. [CrossRef]
29. Wu, Y.; Tang, L.; Huang, C.; Shao, G.; Hou, J.; Sabel, C.E. Enhancing human well-being through cognitive and affective pathways linking landscape sensation to cultural ecosystem services. *Landsc. Ecol.* **2024**, *39*, 175. [CrossRef]
30. Mao, Y.; Xia, T.; Hu, F.; Chen, D.; He, Y.; Bi, X.; Zhang, Y.; Cao, L.; Yan, J.; Hu, J.; et al. The greener the living environment, the better the health? Examining the effects of multiple green exposure metrics on physical activity and health among young students. *Environ. Res.* **2024**, *250*, 118520. [CrossRef]
31. Shao, J.; Qiu, Q.; Qian, Y.; Tang, L. Optimal visual perception in land-use planning and design based on landsenses ecology. *Int. J. Sustain. Dev. World Ecol.* **2020**, *27*, 233–239. [CrossRef]
32. Zheng, T.; Yan, Y.; Lu, H.; Pan, Q.; Zhu, J.; Wang, C.; Zhang, W.; Rong, Y.; Zhan, Y. Visitors' perception based on five physical senses on ecosystem services of urban parks from the perspective of landsenses ecology. *Int. J. Sustain. Dev. World Ecol.* **2020**, *27*, 214–223. [CrossRef]
33. Li, Y.; Li, M.; Xu, Y.; Tao, J. "Interface-element-perception" model to evaluation of urban sidewalk visual landscape in the core area of Beijing. *Front. Arch. Res.* **2024**, *13*, 960–977. [CrossRef]
34. Liu, J.; Yang, L.; Xiong, Y.; Yang, Y. Effects of soundscape perception on visiting experience in a renovated historical block. *Build. Environ.* **2019**, *165*, 106375. [CrossRef]
35. Ba, M.; Kang, J. Perception and Behaviour under the Audio-olfactory Interaction in Urban Public Open Spaces. *South Archit.* **2022**, *10*, 19–29.

36. Asadi-Shekari, Z.; Moeinaddini, M.; Aghaabbasi, M.; Cools, M.; Shah, M.Z. Exploring effective micro-level items for evaluating inclusive walking facilities on urban streets (applied in Johor Bahru, Malaysia). *Sustain. Cities Soc.* **2019**, *49*, 101563. [CrossRef]
37. Hair, J.F.; Risher, J.J.; Sarstedt, M.; Ringle, C.M. When to use and how to report the results of PLS-SEM. *Eur. Bus. Rev.* **2019**, *31*, 2–24. [CrossRef]
38. Hair, J.F.; Sarstedt, M.; Ringle, C.M.; Sharma, P.N.; Liengaard, B.D. Going beyond the untold facts in PLS–SEM and moving forward. *Eur. J. Mark.* **2024**, *58*, 81–106. [CrossRef]
39. Zhang, Y.; Chen, N.; Du, W.; Li, Y.; Zheng, X. Multi-source sensor based urban habitat and resident health sensing: A case study of Wuhan, China. *Build. Environ.* **2021**, *198*, 107883. [CrossRef]
40. White, M.; Smith, A.; Humphryes, K.; Pahl, S.; Snelling, D.; Depledge, M. Blue space: The importance of water for preference, affect, and restorativeness ratings of natural and built scenes. *J. Environ. Psychol.* **2010**, *30*, 482–493. [CrossRef]
41. Curl, A.; Mason, P. Neighbourhood perceptions and older adults' wellbeing: Does walking explain the relationship in deprived urban communities? *Transp. Res. Part A Policy Pract.* **2019**, *123*, 119–129. [CrossRef]
42. Zang, P.; Qiu, H.; Zhang, H.; Chen, K.; Xian, F.; Mi, J.; Guo, H.; Qiu, Y.; Liao, K. The built environment's nonlinear effects on the elderly's propensity to walk. *Front. Ecol. Evol.* **2023**, *11*, 1103140. [CrossRef]
43. Bertram, C.; Rehdanz, K. The role of urban green space for human well-being. *Ecol. Econ.* **2015**, *120*, 139–152. [CrossRef]
44. Lin, W.; Chen, Q.; Jiang, M.; Zhang, X.; Liu, Z.; Tao, J.; Wu, L.; Xu, S.; Kang, Y.; Zeng, Q. The effect of green space behaviour and per capita area in small urban green spaces on psychophysiological responses. *Landsc. Urban Plan.* **2019**, *192*, 103637. [CrossRef]
45. Li, Y.; Yabuki, N.; Fukuda, T. Exploring the association between street built environment and street vitality using deep learning methods. *Sustain. Cities Soc.* **2022**, *79*, 103656. [CrossRef]
46. Vichiensan, V.; Nakamura, K. Walkability Perception in Asian Cities: A Comparative Study in Bangkok and Nagoya. *Sustainability* **2021**, *13*, 6825. [CrossRef]
47. Sun, X.; Wang, L.; Wang, F.; Soltani, S. Behaviors of seniors and impact of spatial form in small-scale public spaces in Chinese old city zones. *Cities* **2020**, *107*, 102894. [CrossRef]
48. Zeng, Q.; Wu, H.; Zhou, L.; Huang, G.; Li, Y.; Dewancker, B.J. Toward pedestrian-friendly cities: Nonlinear and interaction effects of building density on pedestrian volume. *J. Transp. Geogr.* **2024**, *119*, 103954. [CrossRef]
49. Liu, J.; Xiong, Y.; Wang, Y.; Luo, T. Soundscape effects on visiting experience in city park: A case study in Fuzhou, China. *Urban For. Urban Green.* **2018**, *31*, 38–47. [CrossRef]
50. Maculewicz, J.; Erkut, C.; Serafin, S. How can soundscapes affect the preferred walking pace? *Appl. Acoust.* **2016**, *114*, 230–239. [CrossRef]
51. Meihui, B.; Kang, J.; Li, Z. The effects of sounds and food odour on crowd behaviours in urban public open spaces. *Build. Environ.* **2020**, *182*, 107104. [CrossRef]
52. Wang, X.; Tang, P.; He, Y.; Woolley, H.; Hu, X.; Yang, L.; Luo, J. The correlation between children's outdoor activities and community space characteristics: A case study utilizing SOPARC and KDE methods in Chengdu, China. *Cities* **2024**, *150*, 105002. [CrossRef]
53. Yildirim, M.; Globa, A.; Gocer, O.; Brambilla, A. Multisensory nature exposure in the workplace: Exploring the restorative benefits of smell experiences. *Build. Environ.* **2024**, *262*, 111841. [CrossRef]
54. McGann, J.P. Poor human olfaction is a 19th-century myth. *Science* **2017**, *356*, eaam7263. [CrossRef]
55. Xu, J.; Chen, L.; Liu, T.; Wang, T.; Li, M.; Wu, Z. Multi-Sensory Experience and Preferences for Children in an Urban Forest Park: A Case Study of Maofeng Mountain Forest Park in Guangzhou, China. *Forests* **2022**, *13*, 1435. [CrossRef]
56. Hedblom, M.; Gunnarsson, B.; Iravani, B.; Knez, I.; Schaefer, M.; Thorsson, P.; Lundström, J.N. Reduction of physiological stress by urban green space in a multisensory virtual experiment. *Sci. Rep.* **2019**, *9*, 10113. [CrossRef] [PubMed]
57. Zheng, T.; Pan, Q.; Zhang, X.; Wang, C.; Yan, Y.; Van De Voorde, T. Research Note: Linking sensory perceptions with landscape elements through a combined approach based on prior knowledge and machine learning. *Landsc. Urban Plan.* **2024**, *242*, 104928. [CrossRef]
58. Chan, C.-S.; Shek, K.F.; Agapito, D. The sensory experience of visitors with hearing impairment in Hong Kong Wetland Park based on spatial sensory mapping and self-reported textual analysis. *Landsc. Urban Plan.* **2022**, *226*, 104491. [CrossRef]
59. Zhang, T.; Liu, J.; Li, H. Restorative Effects of Multi-Sensory Perception in Urban Green Space: A Case Study of Urban Park in Guangzhou, China. *Int. J. Environ. Res. Public Health* **2019**, *16*, 4943. [CrossRef]

land

MDPI

Article

The Landscape Catalytic Effect of Urban Waterfronts—A Case Study of the Huangpu River in Shanghai

Yuting Yin [1], Dongbo Ma [2] and Xiran Xu [3,*]

1 School of Design, East China Normal University, Shanghai 200062, China; ytyin@design.ecnu.edu.cn
2 Shanghai Tongji Urban Planning and Design Institute, Co., Ltd., Shanghai 200092, China; giser1993mdb@gmail.com
3 College of Landscape Architecture and Art, Henan Agricultural University, Zhengzhou 450002, China
* Correspondence: pinky2580@hotmail.com

Abstract: Waterfronts are some of the most well known public spaces that can catalyse urban changes, yet their benefits have not been systematically explored. This study investigates the potential benefits of waterfront regeneration for the subsequent development of the wider surrounding areas and whether these benefits encompass a broader range of influences. Taking an extensive linear catalyst, the Huangpu River waterfronts in Shanghai, as an example, the catalytic effect of each waterfront section was investigated, visualised, compared and discussed within and across different sections and catalytic influential aspects. A multi-method approach driven by multi-sourced big data was used in this study, and the analysis was carried out at two scales: the waterfront area (the catalyst area) and its surroundings of influence (the areas affected by the catalyst area). The research findings suggest that the landscape catalytic effect is more pronounced in the catalyst area itself than in the surrounding areas affected by the catalyst area. Such effects also vary across waterfront sections, and the western bank of the Huangpu River was more obviously influenced than the eastern bank. The possible reasons for these differences may be related to the area's original function, development limitations and available resources. This study also provides evidence indicating that the relationship between the catalyst and the spatial, social and economic aspects of changes it induces is one of ongoing and mutually supportive interaction. The outcomes of this study include a framework composed of 14 indicators that can disclose the depth and progress of a catalytic effect facilitated by the landscape, as well as implications for the decision-making process in the urban regeneration agenda that centres around waterfronts.

Keywords: urban regeneration; landscape catalytic effect; waterfront; data-driven analysis; spatial mapping

1. Introduction

The catalytic effect of the landscape has been recently emphasised in many Chinese cities to encourage the design of high-quality open spaces as catalysts for facilitating the regeneration of derelict urban areas [1]. Waterfronts and parks are probably the most well known public spaces that can catalyse urban changes [2]. Regenerating urban waterways previously used for industrial purposes has become a worldwide phenomenon to help local communities in terms of social, economic and environmental factors since the 1970s [3]. However, existing literature either assesses the vitality of waterfronts or focuses on the catalytic effect of small-scale punctate projects. To date, no specific attempts have been made to explore the catalytic effect of regenerated urban waterfronts that are characterised by

 https://doi.org/10.3390/land14020422

their linear forms and extensive shorelines. A typical example of this kind is the regenerated Huangpu River waterfront in Shanghai, China. Based on an analysis of the theoretical concept of the landscape catalytic effect and studies of waterfront vitality, this research intends to investigate the landscape catalytic effect of the Huangpu River waterfront and discuss the underlying reasons for bringing about these effects.

1.1. Development and Application of Urban Catalyst Theory

The term 'catalyst' being used in the arena of architecture, urban design and landscape can be traced back to Aldo Rossi's *The Architecture of the City* [4]. Rossi considered that certain primary elements or artefacts in urban settings could bring changes to the surrounding environment not only in the processes of incremental development but also in those of redevelopment [4]. Later, Attoe and Logan defined a catalyst in reference to the 'positive impact an individual urban building, project, program can have on subsequent projects and, ultimately the form of a city' [5] in their book *American Urban Architecture: Catalyst in the Design of Cities*. Similar terms include 'activity generator', 'anchor' [6] and 'magnet' [7], but only when projects facilitate nearby development. The long-term value of catalysts for cities, therefore, is seen not only in openly adapting to the ascribed current and future functions but also in initiatively spurring the dynamics of their surroundings [5]. Urban catalysts are not always 'physical, constructed, measurable artefacts' but can exist in intangible forms (e.g., events, policies and temporal installations) [8]. Oswalt postulated that the temporary use of empty places might promote many changes in the city and introduced the notion of urban catalysts as 'elements or acts of potentially limited duration, initiating processes that may continue long after they have transformed or disappeared' [9]. A catalyst can positively transform existing urban elements without obliterating their value while the catalyst itself remains identifiable rather than being consumed in the catalytic reaction [6]. Based on the characteristics of urban catalysts and the reactions they promote, some scholars have asserted that the essence of urban catalysts partly resonates with Jacob's focus on everyday processes [10], such as Norberg-Schulz's 'genius loci' [11], which states that a city has its own form, spirit, identity and sense of place, and Lynch's assertion that if 'utopian ideas' are to have any chance of success, 'they must be fuller of the substance of life that people know' [12].

A positive catalytic reaction should consider the local context and influence people's perception of the area, create pedestrian traffic by providing a primary destination, visually and physically connect with the surroundings and complement streetscapes [13], increase business and tax revenues by attracting investments, benefit the environment and promote the sustainability of local community [14]. Urban catalysts may vary in their physical shapes, functions and development intentions (e.g., waterfront [15], industrial heritage [16], rail transit [17] and historic preservation [18]), but their success is only determined by whether the redevelopment of their surroundings is ignited. Sternberg [6] suggested that the relationship between a catalyst and its surroundings can be evaluated by using an index, but the categories and weight scale by which the assessment should be conducted were not specified in his work. Some measures of property-led urban regeneration (e.g., the amount of used or reused floor space and the number of jobs created within the surrounding areas) have been employed in related attempts [19]. For example, Chinese scholars Chen and his colleagues adopted activity, modifiability and selective guidance as three indicators for evaluating the catalytic effect of urban complexes [20] but did not provide a corresponding calculation method. Very limited efforts have been made to quantify the catalytic effects of open spaces and landscapes. Only one study carried out in Chongqing established a framework from the dynamic, temporal and spatial perspectives, and incorporated six indicators: user concentration, supplier concentration, landscape topicality, facility correlation, user satisfaction, and matching degree of supply and demand [21]. However, it delineated landscape catalysts based on land use patches rather than the precise boundaries of the catalysts and overlooked the fact that relevant assessments should be conducted on two scales: the catalyst itself and the area it catalyses. Though waterfronts have long been regarded as a catalyst for city renewal [15], previous studies only discussed their role in creating the sense of place [22], shaping urban morphology [23], and providing attractions to local residents and tourists [24] based on case studies. The catalytic effect of waterfronts has not yet been comprehensively explored and quantitatively analysed to provide practical implications for the urban regeneration agenda.

1.2. Urban Waterfront Regeneration in Worldwide Practices

Since ancient times, human civilisations have been born near rivers. The dynamism of rivers represents the collective product of not only geology, ecology and climate but also economy, technology, politics and human perceptions [21]. Urban waterfronts have traditionally served as centres for trade, commerce and transit [21], nd played important roles in the social and economic development of their surrounding communities. From the industrial era to the post-industrial era, many cities have seen the rise of industrialisation along waterways and, more recently, a decline in their banks [3]. In the 1980s, Hall and Jones raised the waterfronts' revitalisation as the major event in urban planning and compared their importance to that of motorways and new residential areas of the 1950s and 1960s [25]. Waterfront regeneration is increasingly popular in this context as the process of transforming former industrial or derelict waterfronts into new, mixed-use areas that promote economic development, environmental sustainability and community vitality [26]. North America is widely recognised as the cradle of waterfront regeneration. Cities such as Baltimore, Boston and New York have set classic examples of renewal for other cities with waterfront areas [1,3,27]. Since the 1970s, the three major British port cities—London, Liverpool and Bristol—have achieved significant transformation of their abandoned docklands [28]. This has been followed by other cities such as Cardiff, Dundee, Edinburgh, Glasgow and Swansea. The movement of waterfront redevelopment in continental European countries began in the 1980s [29]. Barcelona [30], Amsterdam and Berlin have also been actively involved in this trend [29]. Famous examples also include the Darling Harbour in Sydney [31] and projects conducted in Asian port cities such as Dubai in Abu Dhabi, Hong Kong and Shanghai in China, Tokyo in Japan and Singapore [32].

These waterfront renewal projects have contributed to improve connectivity between the water and the city; generate new economies and dynamics of regional development [33]; introduce complementary facilities that serve the city in a multiscale way; renew urban green corridors or regenerate heritage features; and create landmarks beyond the city waterfronts [30,34]. Waterfront regeneration is typically a long-term process, with its focus varying according to the local context and development objectives. Therefore, the success of waterfront regeneration is usually evaluated from multiple perspectives. However, to date, there has been no exploration into the impact of the landscape catalytic effect on waterfront regeneration from the perspective of planning and design. The closest related research arena is on the vitality of waterfront areas, since vitality has been regarded as the most important criterion for determining whether its value as open spaces has been realised. The theoretical basis of open space vitality was formed by many famous urban researchers, including Jacobs [10], Lynch [11] and Gehl [35], who named people, activities and space as three fundamental elements [36]. Later, Mehta [37] added that vitality can be reflected in interactions between human activities and physical environments, as it requires large numbers of people to be involved in a series of fixed or continuous activities, especially social ones. Accordingly, the existing literature primarily revolves around the three elements and either evaluates the vitality level of waterfront spaces or explores the environmental factors that affect their vitality.

The relevant explorations relied on questionnaires, interviews, observations and behaviour mapping to reveal people's gathering characteristics (e.g., intensity, duration, frequency and with or without companies), movements and preferred activities (e.g., exercising, socialising and sightseeing), and spaces that allow these activities to indicate the level of vitality [38–41]. The coming of the data era has provided urban planners and designers with more insights into whether the vitality goal has been achieved. Heat map data and mobile signalling data provided by open-sourced platforms are commonly used to visually represent the density and distribution of a population during different periods [42]. With their help, the vitality along the Huangpu River in Shanghai [43], Yangtze River in Nanjing [44] and Shenzhen Bay waterfront area [45] has been measured with different mathematic models. These vitality evaluation results have then been correlated with potential environmental factors to identify those with significant influences. For the waterfront itself, service facilities, shoreline curvature [46], site design [47], the spatial type of waterfront area and visual openness [48] are dominant factors influencing vitality, while external factors may include road network density, slow lane accessibility, green space rate and building density [49]. However, the assessment of waterfront vitality does not examine the catalytic effect brought about by regeneration, as it only indicates the intensity and diversity of human activities occurring within a waterfront area and the environmental cues that may be responsible for them. The catalytic effect involves more complicated outcomes–specifically, whether the regeneration of a waterfront area can benefit the subsequent development of the wider surrounding area and whether these benefits encompass a broader range of influences.

1.3. The Aim of This Study

Driven by multi-sourced big data, this study intended to explore the potential catalytic effects that waterfront regeneration can bring to its surrounding communities. Taking the regeneration of the Huangpu River waterfronts in Shanghai as an example, this study first identified indicators from the overlaps between waterfront vitality research and urban catalyst studies to describe the positive effects of waterfront regeneration. Next, the weight scale of each indicator was determined using the analytic hierarchy process (AHP). The catalytic effect of each waterfront section along the Huangpu River was then investigated, visualised, compared and discussed within and across different sections and different catalytic influential aspects. This analysis was carried out at two scales: the waterfront area and its surroundings of influence.

2. Materials and Methods

2.1. Research Sites of the Huangpu River in Shanghai, China

One of the well-known landscape catalysts in Shanghai is the regenerated waterfront along the Huangpu River. It was used by manufacturing industries, such as shipbuilding, waterworks, cement making and aircraft making. As these industries relocated, the banks of the Huangpu River (hereinafter also referred to as the river) in the core area of Shanghai progressively declined. From 2016 to 2017, Shanghai carried out the Huangpu River Core Section Connectivity Project to transform its 45 km waterfront into public spaces. This not only formed a connected linear open space in a densely populated urban territory but also encouraged the proliferation of commerce, and cultural and economic industries, as well as the construction of residential apartments, which later became some of the most expensive in the city.

Within this context, this study selected the core section of the Huangpu River waterfront delineated in the 'Development Plan for Huangpu River and Suzhou River' as the research site. The site covers approximately 45 km of shoreline between Yangpu Bridge and Xupu Bridge, with a depth of about 0.5 km on each side of the riverbank as the catalyst itself and a depth of about 2 km as the area affected by the catalyst. The total size of the catalyst area (CA) is about 12.00 km^2, and the area affected by the catalyst area (CAA) is around 46.04 km^2 (see Figure 1).

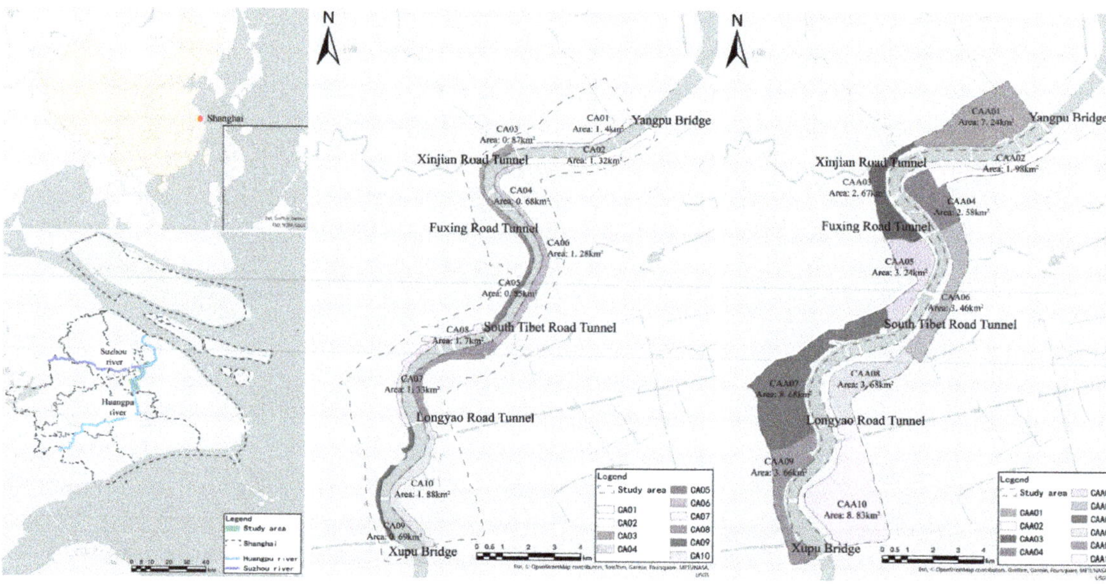

Figure 1. Research sites of the catalyst and its influenced areas.

For the comparative analysis, the research site was further divided into 20 areas by referring to previous similar studies [50]: CA01/CAA01 and CA02/CAA02 are the areas on both sides of the interval between Yangpu Bridge and the Xinjian Road Tunnel; CA03/CAA03 and CA04/CAA04 are the areas on both sides of the interval from the Xinjian Road Tunnel to the Fuxing Road Tunnel; CA05/CAA05 and CA06/CAA06 are the areas on both sides of the interval from the Fuxing Road Tunnel to the South Tibet Road Tunnel; CA07/CAA07 and CA08/CAA08 are located between the South Tibet Road Tunnel and the Longyao Road Tunnel; and CA09/CAA09 and CA10/CAA10 refer to the areas on both sides of the interval from the Longyao Road Tunnel to the Xupu Bridge (see Figure 1).

2.2. Indicators for Disclosing the Catalytic Effects of Huangpu River Waterfront

Most extant literature concludes that the urban catalyst effect is a complex process that involves spatial, social and economic changes to indicate its activity, characteristics and inertia [2,8,13,21]. The environmental attributes affecting the vitality of public spaces include natural ecology, spatial function, historical and cultural value, accessibility, surrounding development and policy management [37,39–43]. Indicators for disclosing waterfront catalytic effects were selected from the overlaps between existing urban catalyst studies and waterfront vitality studies based on the following criteria: (1) indicators should be closely in response to local contexts of the river; (2) indicators should be able to present the effects occurring within the CAs and CAAs; (3) the selected indicators can be described by quantitative data; (4) data of selected indicators can be obtained online, and the years of the most recent data should be consistent; and (5) data of selected indicators should cover the research site as much as possible.

Considering waterfront features, data availability and the requirements for quantitative calculation, this study incorporated 14 indicators to describe the prominent consequences of waterfront catalytic effects in the spatial, social and economic aspects (Table 1).

Table 1. Identified indicators for describing waterfront catalytic effects.

	Indicators	Definitions	Data Sources	References
Spatial effect	Building density	The average ratio of the total floor area to the street block area in the unit area	Building height of Asia in 3D-GloBFP https://zenodo.org/records/11397015 (accessed on 13 October 2024)	[49,51–54]
	Road network density	The length of the road network to the unit area	OpenStreetMap https://www.openstreetmap.org/ (accessed on 5 November 2024)	
	Metro station service acreage	The ratio of the 1200 m radius area of rail stations within the research site to the unit area	Amap https://www.amap.com/ (accessed on 25 October 2024)	
	Bus station service acreage	The ratio of the 1200 m radius area of bus stations within the research site to the unit area		
	Normalised difference vegetation index (NDVI)	The degree of vegetation coverage within the unit area	United States Geological Survey https://earthexplorer.usgs.gov/ (accessed on 15 October 2024)	[55]
Social effect	Parks	The ratio of the 500 m radius area of parks within the research site to the unit area	Amap https://www.amap.com/ (accessed on 25 October 2024)	[45,47,53]
	Public service facilities	The proportion of the number of public service facilities (e.g., relay stations, toilets and parking lots) to the unit area		
	Historical and cultural facilities	The ratio of the number of historical and cultural facilities (e.g., art museums, science museums, memorial halls and exhibition centres) to the unit area		
	User concentration	The human activity level reflected by the night light index in the unit area	Luojia-1A Satellite http://59.175.109.173:8888/index.html (accessed on 1 November 2024)	
	Social media profile	The network influence of the research site on social media reflected by the Weibo search index	Sina Weibo https://s.weibo.com/ (accessed on 12 October 2024)	
Economic effect	Development density	The proportion of constructed or developed areas to the unit area	National statistics https://zenodo.org/record/8176941 (accessed on 19 October 2024)	[36,56–58]
	Functional diversity	The ratio of the points of interest in five categories: leisure and entertainment, business offices, hotel accommodations, sports and fitness, and restaurants within the unit area	Amap https://www.amap.com/ (accessed on 25 October 2024)	
	Rent	The daily rental price per square metre of office buildings in the unit area	Lianjia http://www.lianjia.com/ (accessed on 13 October 2024)	
	Permanent population density	The ratio of the total population in its subordinate street neighbourhood to the area of the street neighbourhood	National statistics https://tjj.sh.gov.cn (accessed on 19 October 2024)	

Note: service radii are determined referring to Code for Transport Planning on Urban Road GB 50200-1995 [59] and Code for the Design of Public Park GB51192-2016 [60].

2.3. Data Collection and Analysis

2.3.1. Collection and Calculation of the 14 Catalytic Indicators

Data on the 14 indicators were retrieved from national statistics and online open-sourced datasets. To ensure that all data were from the same time and up to date, the year 2021 was set as the timestamp for extracting all relevant data. The calculation formulas refer to previous similar studies and are appropriately adjusted according to the specific circumstances of this research (Table 1).

Four indicators in the spatial aspect—building density, road network density, metro and bus station service acreage—were calculated based on the total acreage of the area and defined as [refer to Equations (1)–(3)]:

$$FAR = Agf \div A \tag{1}$$

where

FAR: the building density of the area;

Agf: the total floor acreage of the buildings in each area;

A: the total acreage of the area.

$$D = L \div A \tag{2}$$

where

D: the road network density;

L: the total length of the road network, including the trunk roads, collector roads and branch roads;

A: the total land acreage of the area.

$$S_m = S_b = \sum_1^n \pi {r_1}^2 \div A \tag{3}$$

where

S_m: the service acreage of the metro stations;

S_b: the service acreage of the bus stations;

n: the number of metro or bus stations in each area;

r_1: the service radius (1200 m) of each metro or bus station;

A: the total land acreage of the area.

In terms of the social aspect, three indicators—parks, public service facilities, historical and cultural facilities—were defined as [refer to Equations (4)–(6)]:

$$S_p = \sum_1^n \pi {r_2}^2 \div A \tag{4}$$

where

S_p: the service acreage of the parks;

n: the number of parks in each area;

r_2: the service radius (500 m) of each park;

A: the total land acreage of the area.

$$S_{pf} = F_i \div A \tag{5}$$

where

S_{pf}: the service level of the public facilities;

F_i: the number of public service facilities;

A: the total land acreage of the area.

$$S_{hc} = R_i \div A \tag{6}$$

where

S_{hc}: the service level of the historical and cultural facilities;

R_i: the number of historical and cultural facilities;

A: the total land acreage of the area.

Development density and functional diversity in the economic aspect could not be directly obtained and, thus, were defined as [refer to Equations (7) and (8)]:

$$DD = Aba \div A \quad (7)$$

where

DD: the development density of the area;
Aba: the vertical projection acreage of the constructed buildings in the area;
A: the total acreage of the area.

$$FM = -\sum_{i=1}^{n} P_i \log(P_i) \quad (8)$$

where

FM: the functional diversity;
n: the number of points of interest (PoIs) categories within an area;
$i = 1, \ldots, 5$;
P_i: the proportion of the i type of PoIs.

2.3.2. Indicator Weights Determined Using the Analytic Hierarchy Process (AHP) Method

This study employed the AHP method to determine the importance of the 14 indicators of the landscape catalytic effect. The AHP, developed by Saaty in 1987, is an effective method to help determine the hierarchy and importance of indicators in a blurring situation [61]. It has been widely used in landscape studies in relation to evaluation [62,63], management decision-making [64,65] and spatial optimisation [66,67]. Ten experts in the fields of architecture, planning and landscape design were invited to compare the importance of these indicators by scoring them in pairwise comparisons through an online questionnaire. Three levels in the AHP were determined before analysis: the landscape catalytic effect was the destination level; spatial, social and economic were referred to as the criterion level; and the 14 indicators belonged to the index level (Figure 2).

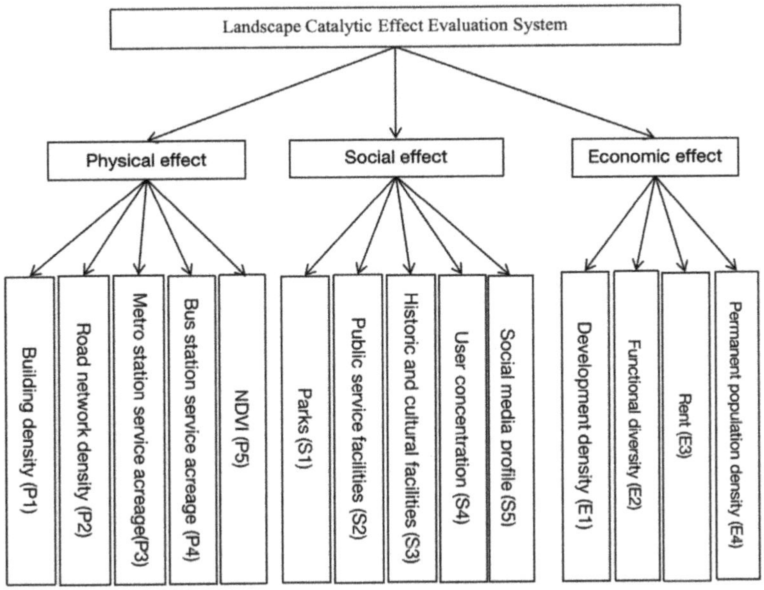

Figure 2. AHP hierarchical structure of landscape catalytic effect.

The procedure for conducting the AHP was as follows. (1) A judgment matrix was constructed for each of the 10 experts. (2) The average values of these matrices were used to generate the final judgment matrix of the criterion level [refer to Equation (9)] using the average scores of the 10 experts, and was normalised by columns [refer to Equation (10)]. The characteristic vector of the criterion level was calculated [refer to Equations (11) and (12)]. (3) The matrix was subjected to consistency tests [refer to Equations (13)–(15)]. The weights of the three factors in the criterion level could only be calculated if the consistency results were reasonable. (4) The judgment matrix and weights of the indicators within each criterion level were outputted. (5) The comprehensive weights of the 14 indicators could then be obtained by multiplying the weights of the criterion level with those of the indicator level (See supplementary material for the complete procedure of AHP).

$$r_i = \sum_{j=1}^{n} r_{ij} i = 1, 2, \ldots, n \quad (9)$$

$$b_{ij} = \frac{r_{ij}}{\sum_{i=1}^{n} r_{ij}} i, j = 1, 2, \ldots, n \quad (10)$$

where

b_{ij}: the element in the ith row and jth column of the normalised matrix;
r_i: the sum of the elements in the ith row of the judgment matrix;
r_{ij}: the element in the ith row and jth column of the judgement matrix;
n: the number of elements in the matrix;
i: the ith row;
j: the jth column.

$$\overline{W} = \sum b_{ij} i, j = 1, 2, \ldots, n \quad (11)$$

$$w_i = \frac{\overline{W_i}}{\sum_{i=1}^{n} \overline{W_i}} i = 1, 2, \ldots, n \quad (12)$$

where

$\overline{W}$: the characteristic vector;
$\overline{W_i}$: the ith element in the characteristic vector $\overline{W}$;
w_i: the weight value of the ith indicator;
n: the number of elements in the matrix;
i: the ith row.

$$\lambda_{max} = \frac{1}{n} \sum_{i=1}^{n} \frac{\left(B \times \overline{W}\right)_i}{w_i} \quad (13)$$

$$CI = \frac{\lambda_{max} - n}{n - 1} \quad (14)$$

$$CR = \frac{CI}{RI} \quad (15)$$

where

λ_{max}: the maximum eigenvalue;
n: the number of elements in the matrix;
B: the judgement matrix;
CI: the consistency indicator;
RI: the proportionality coefficient;
CR: the random consistency ratio.

2.3.3. Visualisation Using ArcGIS Platform

The calculation results of the eight indicators listed in Section 2.3.1 were entered into the ArcGIS 10.8, together with the five other indicators that could be directly retrieved, for processing and visualisation. Mapping results were provided for each of the 14 indicators, the three catalytic aspects (spatial, social and economic) and the overall landscape catalytic effect. The latter two were calculated based on the weights of the three aspects and the 14 indicators, respectively. All the visualisations were taken at two levels: CA and CAA.

3. Results

3.1. Mapping Results of the 14 Landscape Catalytic Indicators

Five indicators in the spatial aspect describing the landscape catalytic effect of the research site appeared in three different patterns in the mapping results. First, the building density and road network density were relatively lower in the CAs than in the CAAs. Among the 10 CAAs, CAA03 and CAA04 had the densest buildings and road networks, followed by CAA05, CAA06 and CAA01, while CAA10 had the lowest density. The distribution characteristics were in line with those in the CAs, and relatively lower densities were observed in CA07 and CA09, but not CA10. The metro and bus station services shared the second distribution pattern, and their mapping results were very much alike. Nine CAs (excluding CA10) and nine CAAs (excluding CAA10) were at the same level of metro station service (0.92–1.00) and bus station service (0.99–1.00). CA10 and its affected area (CAA10) were at the lowest level in terms of metro and bus station services, with CAA10 slighter lower than CA10. Differing from the other four indicators, the normalised difference vegetation index (NDVI) was observed to be higher in the CAs compared with the CAAs. The highest level of the NDVI in the CAs was found in CA08 and CA10, comprising Qiantan Park and Shanghai Expo Park, while CAA04, where Oriental Pearl TV Tower Park and the Lujiazui Central Green Space are located, had the highest NDVI among the CAAs (see Figure 3).

The mapping results of the indicators in the social aspect disclosed both similarities and differences. Overall, no obvious differences existed between the CAs and the CAAs in the distribution of parks, public service facilities, historical and cultural facilities, and user concentration, but the CAs had higher social media profiles compared with the CAAs. Public service facilities and historical and cultural facilities were mostly concentrated in CA03 and CAA03, but the former had seven areas (CA01, CAA01, CAA 02, CAA04, CA05, CAA05 and CAA07) ranked at the second level, while the latter only had three (CA01, CAA04 and CA07). In general, public service facilities were more abundant than historical and cultural facilities within the research site. The distribution of parks within the site was very close to the characteristics of the public service facilities and only differed in the highest areas, with CA04, CA004 and CA05 having the densest level of parks (0.74–0.85). The user concentration results suggested that the CAAs between the Xinjian Road Tunnel and the South Tibet Road Tunnel (CAA03, CAA04 and CAA05) were the most popular. The eastern bank of the Huangpu River between Yangpu Bridge and the Xinjian Road Tunnel (CA02 and CAA02) also attracted plenty of people. The southern part of the research site, especially the areas close to Xupu Bridge (CA09 and CA10), had fewer users than the middle sections. The social media profile mapping revealed that people were most interested in CA07, where the West Bund Museum and the Oil Tank Art Center are located, and CAA10, where Qiantan Park is located. The areas located along both sides of the Huangpu River between Yangpu Bridge and the Xinjian Road Tunnel (CAA01, CAA02) and CA04 between the Xinjian Road Tunnel and the Fuxin Road Tunnel were the least attractive (see Figure 4).

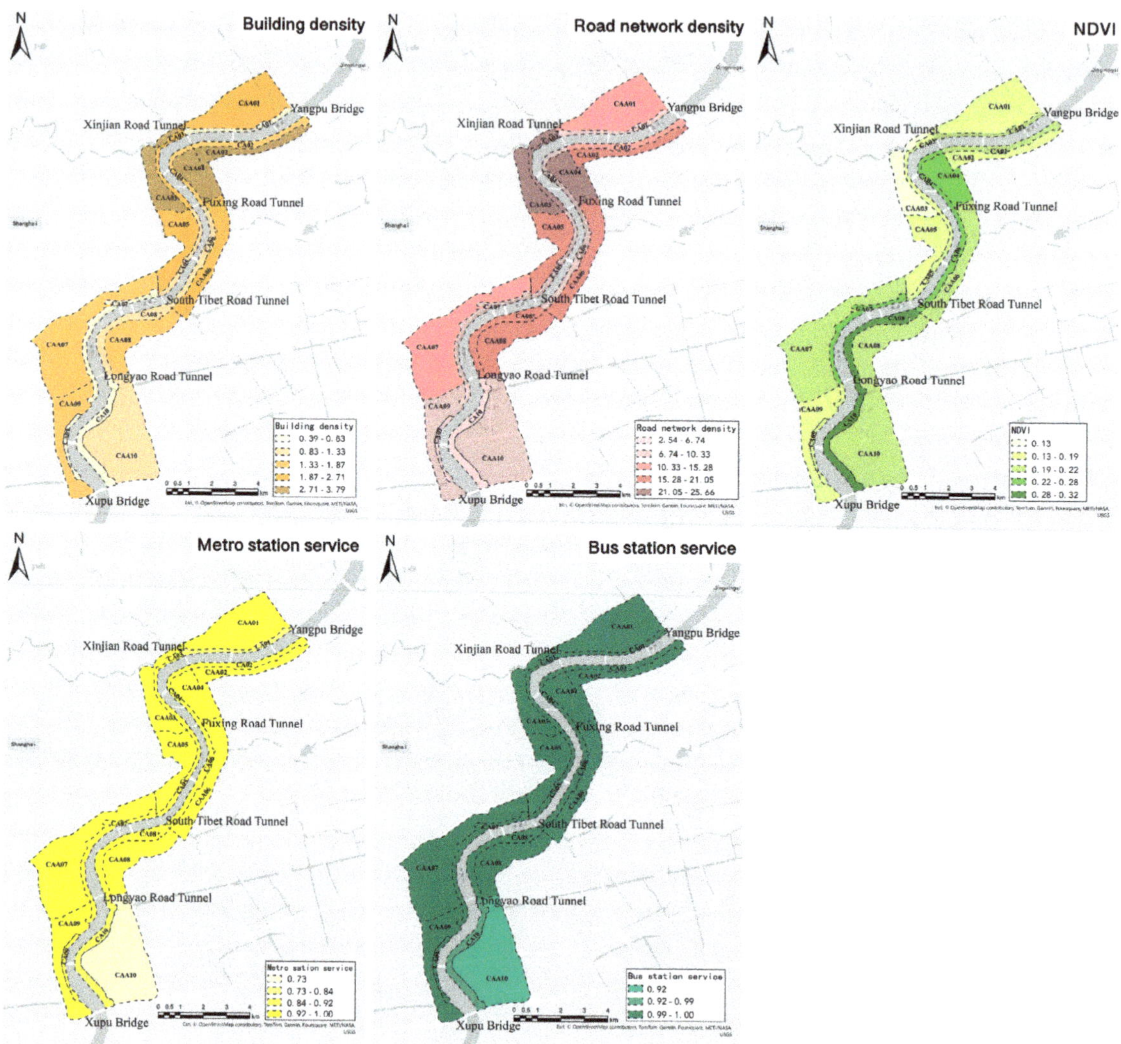

Figure 3. Mapping results of the five indicators in the spatial aspect.

The distribution characteristics of the four indicators in the economic aspect were also different from each other. The development densities of almost all the CAAs within the research site were at the highest level (0.66–0.73), apart from CAA10 (0.47–0.61), and were generally higher than the CAs. Among the 10 CAs, only CA08 reached the range of 0.66–0.73. CA09 had the lowest development density (0.37). Functional diversity was highest in the waterfront section between the Fuxing Road Tunnel and the South Tibet Road Tunnel (CA05, CAA05, CA06 and CAA06), but its two adjacent sections (Xinjian Road Tunnel–Fuxing Road Tunnel and South Tibet Road Tunnel–Longyao Road Tunnel) were less diverse, especially the 04 and 08 areas. Rental prices were evenly distributed in the five CAAs on the western bank of the Huangpu River, while on the eastern bank the buildings in CAA04 had the highest rental prices. Most CAAs were restricted to commercial development; hence, only CA03 (North Bund) and CA06 (Bund SOHO) had office buildings, with rents ranking the second highest. The population was mainly concentrated in the

CAAs, especially along both sides of the riverbanks between the Xinjian Road Tunnel and the South Tibet Road Tunnel (CA03, CAA03, CAA04, CA05, CAA05 and CA06). These areas are some of the most densely populated residential areas in the central urban district of Shanghai, with most residential communities being established earlier. CAA01 and CAA02 in the north end of the site also had relatively higher levels of permanent population density (269.13–308.21), while the four areas (CA09, CAA09, CA10 and CAA10) in the south end of the site (Longyao Road Tunnel–Xupu Bridge) had the lowest population density (67.84–103.31) (see Figure 5).

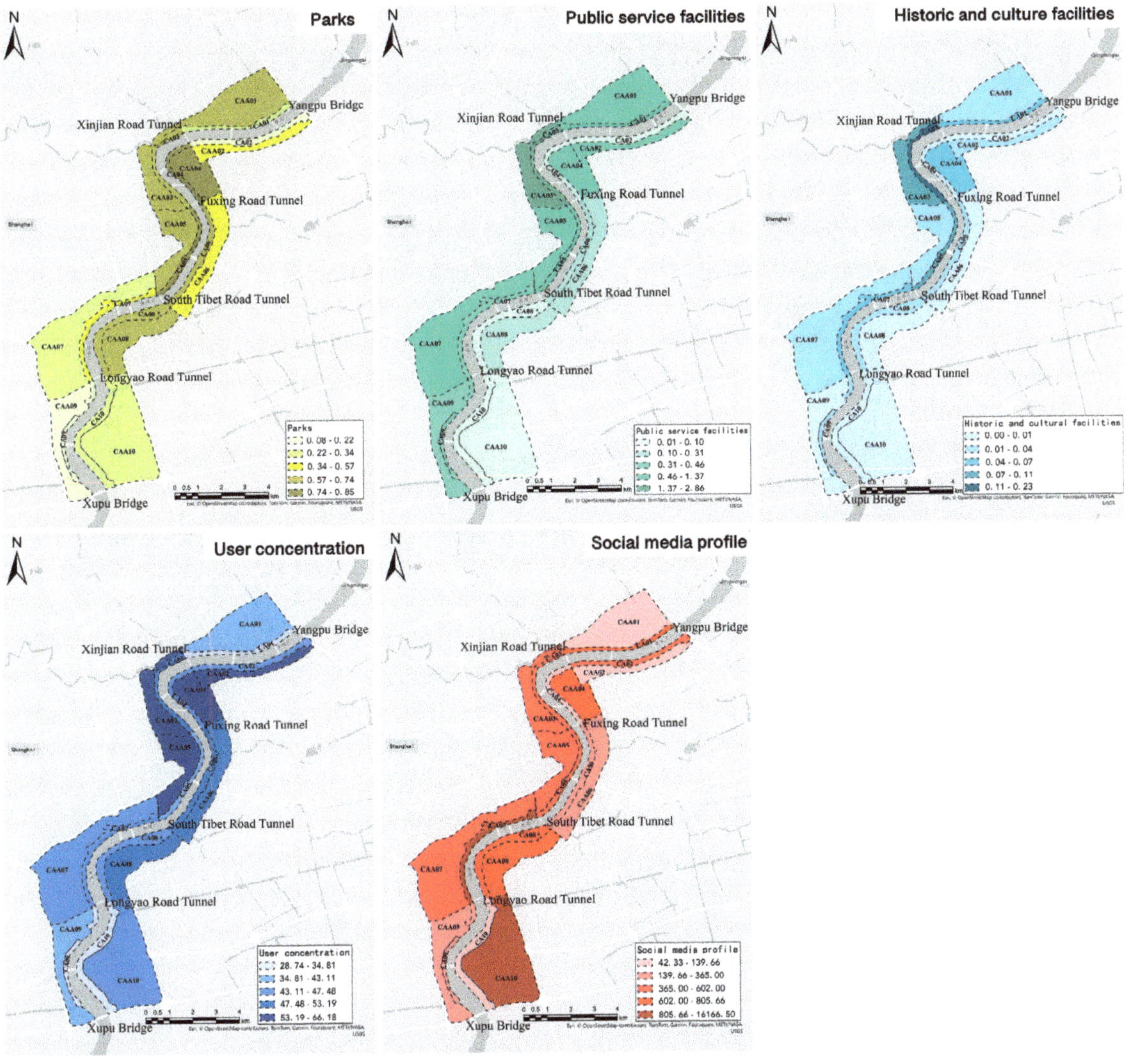

Figure 4. Mapping results of the five indicators in the social aspect.

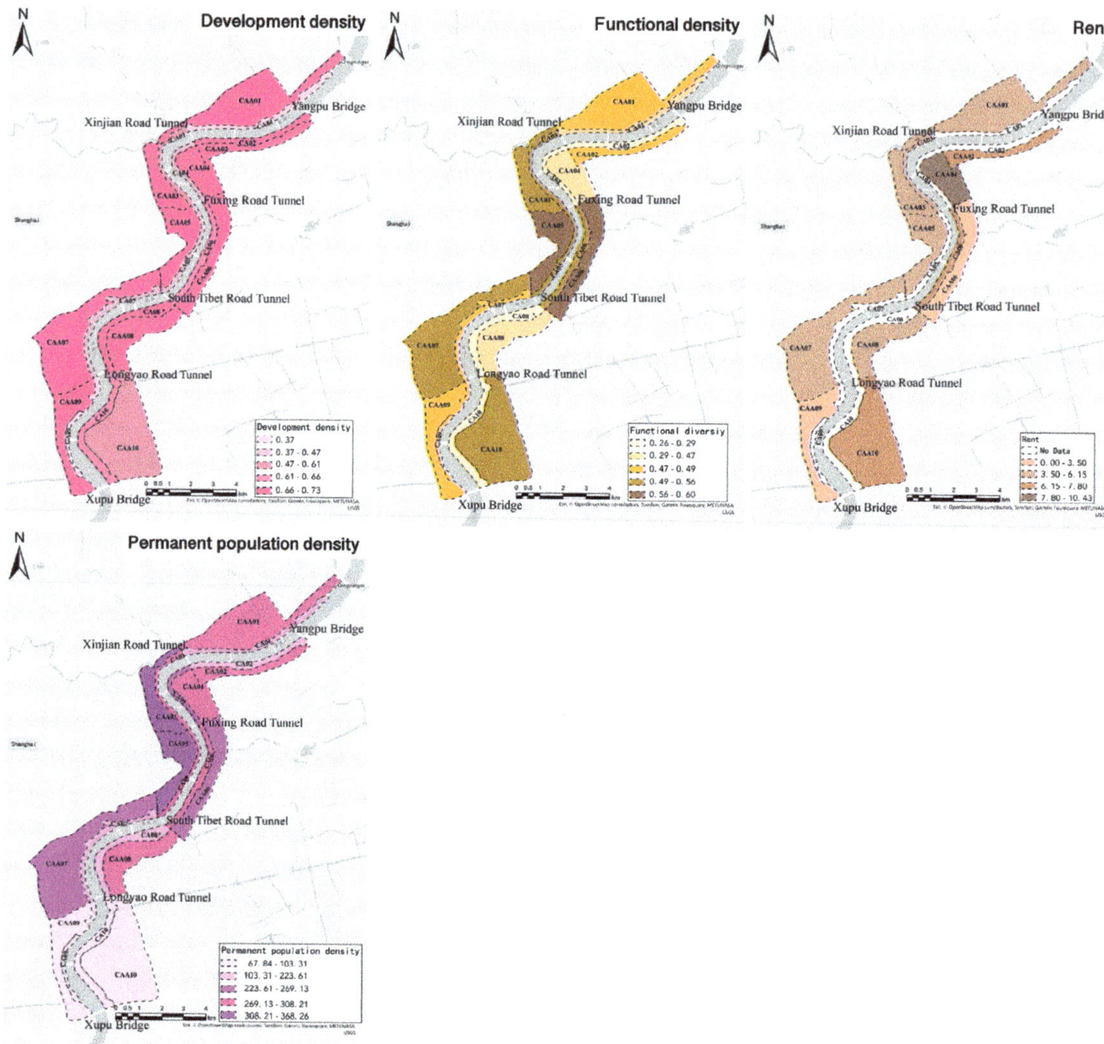

Figure 5. Mapping results of the four indicators in the economic aspect.

3.2. Weight Calculation of the 14 Landscape Catalytic Indicators

The results suggested sufficient consistency ($CR < 0.1$); thus, all the judgement matrixes passed the consistency test. The weights of the three aspects and 14 indicators were calculated, and the comprehensive weights were multiplied accordingly. The results showed that, among the three aspects, the spatial aspect was the most important (0.61), followed by the social aspect (0.28) and the economic effect (0.11). The gap between the spatial aspect and the other two aspects was relatively large. In terms of the 14 indicators, those with higher weight scales appeared evenly within the three aspects, such as building density (0.33) and road network density (0.29) in the spatial aspect, parks (0.45) and public service facilities (0.22) in the social aspect, and development density (0.37) and functional density (0.29) in the economic aspect. When the weights of the criterion levels were considered, the comprehensive weight was highest for building density (0.20), followed by road network density (0.18), parks (0.13) and metro station service acreage (0.12). Social media profile (0.02), permanent population density (0.02) and rent (0.02) were rated as the lowest three (see Table 2).

Table 2. Weighted scales of the 14 indicators.

Destination Levels	Criterion Levels	Weights	Index Levels	Weights	Comprehensive Weights
Landscape catalytic effect	Spatial aspect	0.61	Building density	0.33	0.20
			Road network density	0.29	0.18
			Metro station service acreage	0.20	0.12
			Bus station service acreage	0.08	0.05
			Normalised difference vegetation index	0.10	0.06
	Social aspect	0.28	Parks	0.45	0.13
			Public service facilities	0.22	0.06
			Historic and cultural facilities	0.16	0.04
			User concentration	0.09	0.03
			Social media profile	0.08	0.02
	Economic aspect	0.11	Development density	0.37	0.04
			Functional diversity	0.29	0.03
			Rent	0.16	0.02
			Permanent population density	0.18	0.02

3.3. Mapping Results of the Landscape Catalytic Effect

The spatial, social and economic aspects, and overall landscape catalytic effect were calculated using the weights in Table 2 and visualised using the ArcGIS platform. Generally, the catalytic effects were more evident in the CAAs than in the CAs. The only exception was CA07, which contained many parks and historical and cultural facilities, leading to a higher score in the social aspect, as well as in the overall landscape catalytic effect.

The mapping results indicated that the spatial, social and economic aspects were fundamentally different, with the only similarity being that the areas on the western bank of the Huangpu River between the Xinjian Road Tunnel and the South Tibet Road Tunnel were significantly affected by the regeneration in all three aspects. The spatial effect caused by the development of the Huangpu River was most evident on the western riverbank between the Xinjian Road Tunnel and the South Tibet Road Tunnel (CA03, CAA03 and CAA 05) and the eastern riverbank between Yangpu Bridge and the Fuxing Road Tunnel (CAA02 and CAA04). Two CAs at the south end (CAA09 and CAA10) and CA01 at the north end of the site were less influenced in terms of spatial changes. As for the social effect, it was more evident in the southern part than in the northern part. The most influential areas were CA07 and CAA10, followed by the areas between the South Tibet Road Tunnel and the Longyao Road Tunnel (CAA07, CA08 and CAA08). CAA01 and CAA02, located at the north end of the site, were least affected in the social aspect. Similar to the spatial aspect result, CAA03 and CAA05 had the most obvious economic effects, and the areas adjacent to them on the north side (CAA01) and south side (CAA 07) and opposite them (CA06 and CAA07) ranked at the second level. Economic influences were least observed in areas located at the south end of the site (CA09, CAA09, CA10 and CAA10) (see Figure 6).

Principally, the landscape catalytic effect was most significant in CA07 (358.48) and CAA10 (199.33), while the least affected areas were CA10 (4.90), CAA09 (10.88) and CA09 (11.23) next to Xupu Bridge and CA04 (11.57) between the Xinjian Road Tunnel and the Fuxing Road Tunnel. The calculation results suggested that the degrees of impact from the catalyst were different, with the most reaching 358.48 and the least only 4.9. Comparing the CAs and CAAs, the landscape catalytic effects were more evident in the CAs (Mean = 51.31) than in the CAAs (Mean = 38.10). This tendency was most obvious in three pairs, namely CA01/CAA01, CA02/CAA02 and CA07/CAA07. Three pairs of areas had similar landscape catalytic effects: CA05/CAA05, CA06/CAA06 and CA09/CAA09. Only in the pairs of 03, 04, 08 and 10 were the CAAs more influenced than the CAs; however, the differences were slight. In addition, the western bank (52.75) of the Huangpu River was more obviously influenced than the eastern bank (45.09) (see Figure 6).

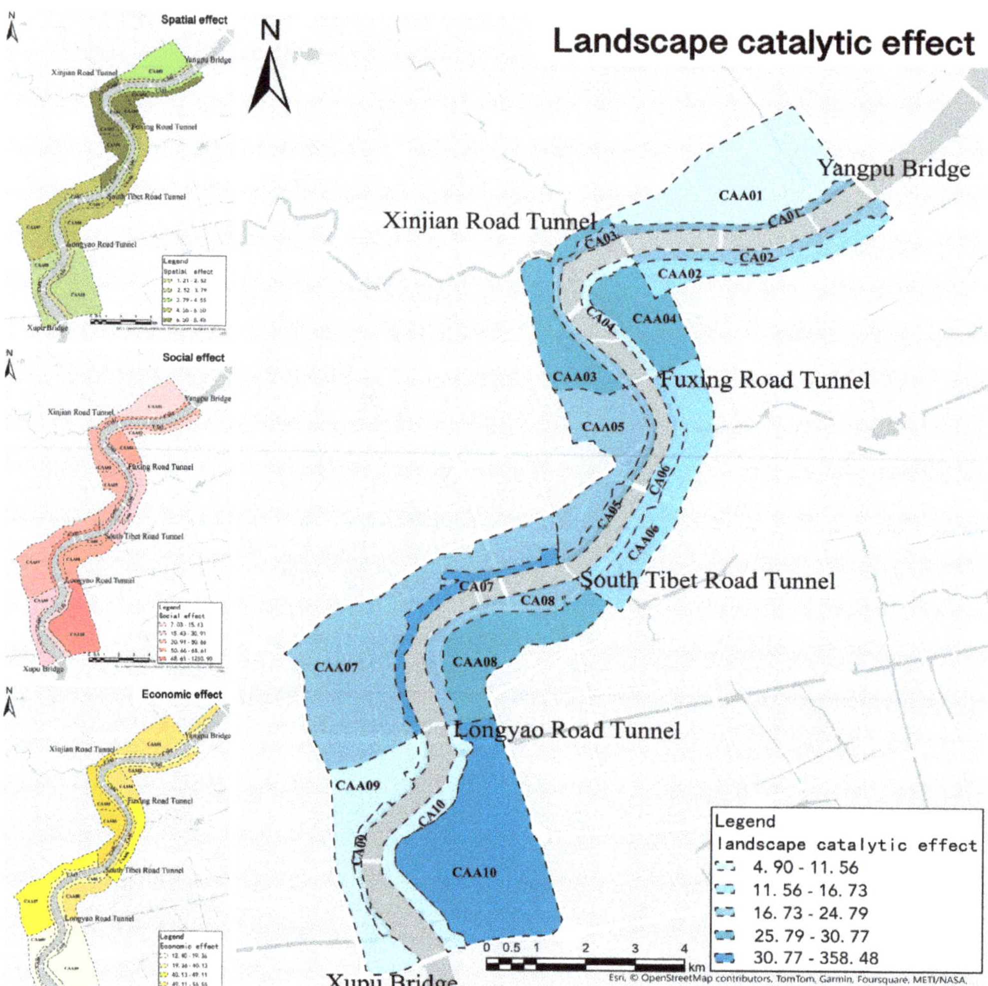

Figure 6. Mapping of the three aspects and the overall landscape catalytic effect.

4. Discussion

4.1. Landscape Catalytic Effect Brought About by Huangpu River Waterfront Regeneration

This study attempted to understand the changes brought about by the landscape catalyst effect of Huangpu River waterfront regeneration. The spatial, social and economic aspects were explored to describe the fundamentals of the catalytic effect. Jacobs stated that 'once one thinks about city processes, it follows that one must think of catalysts to those processes, as this too is of the essence' [10]. The conventional process of a catalytic effect starts with an element (e.g., stadium, theatre and park), a walk-only precinct [68] or a temporal event [4] that can help sustain surrounding retail businesses so that they can, in turn, create enough traffic flow to benefit the operations of the catalyst. Together, they form the 'critical mass'. This is in line with the research findings that the CAs (CA03, CA05, CA06) and CAAs (CAA03, CAA05, CAA06) with higher levels of user concentration and permanent population density within the research site were also those with higher building density, road network density and functional diversity. Although the findings in this study did not show the chronological order in which these changes occur, similar distribution patterns among these indicators also provide compelling evidence regarding the interplay between the catalyst and the changes it induces.

Finding out which attributes of a catalyst facilitate urban changes is essential, as the catalyst itself does not create 'critical mass (a vibrant urban area arises from the interrelationships among multiple buildings)' [69] out of nothing. Sometimes, the composition of a catalyst is as important as the as attributes of it. This is especially true when a vast area such as the Huangpu River serves as a catalyst in its entirety, and it also contains multiple small-scale catalytic elements operating within it—for example, Shanghai Expo Park in CA08 and CA008. Expo events always bring changes to cities by introducing a new cultural taste, expanding areas to organise the expo and attracting new functions [70]. Other examples include the Long Museum in CA07 and the Oil Tank Art Center and the West Bund Museum in CA09. The catalytic effect of cultural and art facilities, such as museums, theatres and stadiums, have greatly impacted cities since the end of the 20th century by attracting a sufficient flow of people to sustain the surrounding retail shops (e.g., pubs, shops and cafes) [71]. The Bund SOHO in CAA05 and the Old Wharf Creative Park in CA05 replaced the old industrial factories with advanced industries of the modern era, and sub-centres grew around them [16]. These catalysts within a catalyst may also have unexpected influences beyond the original landscape catalytic effect (Figure 7).

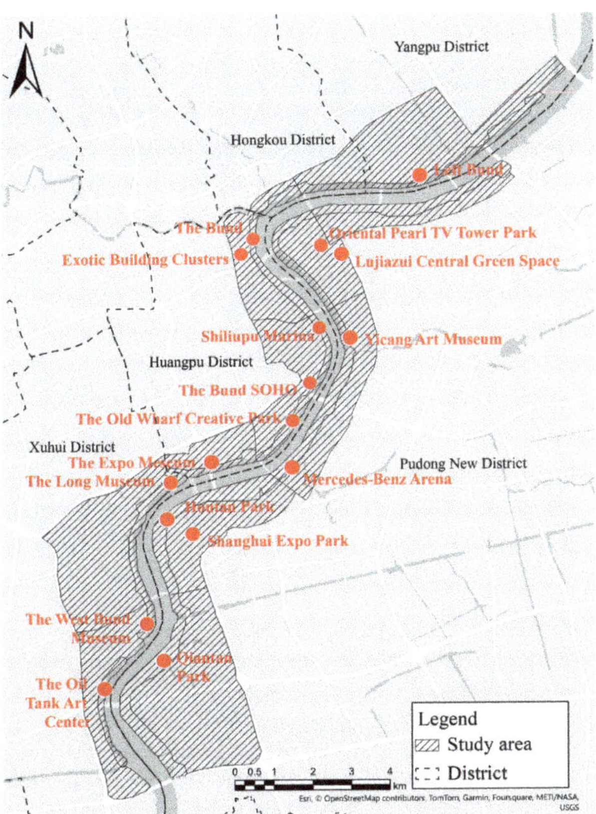

Figure 7. Potential catalysts within the catalyst of Huangpu River waterfronts.

Considering that the catalytic effect is a long-lasting, complex and ongoing process, other dimensions may be used to indicate the success of the catalyst effect, such as its influence as an inspiration for future projects [20], its looseness for postulating an idea of temporary use [8,9], its influence on planning and management policies and its effect on facilitating bottom-up changes [72–74]. Notably, although these aspects for measuring the landscape catalytic effect were proposed in this research, they cannot always be used to define the success of a catalyst due to the open-ended nature of this process.

4.2. Reflecting on the Landscape Catalytic Effect of Local Development Along the River

The Huangpu River waterfront regeneration covers an extensive area, including five administrative districts (Yangpu District, Huangpu District, Hongkou District, Xuhui District and Pudong New District) in Shanghai with different development characteristics and focuses. Thus, the waterfront generates various catalytic processes in these districts and appears with different spatial, social and economic effects.

Yangpu District is one of the earliest developed areas in Shanghai's central city and has limited land resources. Most of the areas in CA01 and CA001 in Yangpu District were already occupied by residential and commercial buildings before regeneration. The waterfront regeneration significantly improved the overall environmental quality of the region (spatial) and drove up the vitality of local businesses (economic). However, the catalytic effect in this area was not significantly reflected in its social aspect due to the scarce land resources available for the construction of new cultural and historical buildings and urban green spaces.

CA09 and CA009 in Xuhui District also had less obvious catalytic effects, but their weakest aspect was economic development. These areas are mostly composed of old neighbourhoods built around 30 years ago and office buildings recently constructed in the original industrial areas after the regeneration of the Huangpu River. Hence, the landscape catalytic effect may have lagged in this section, leading to a situation where, although the development intensity within the area has been sufficient, it has not attracted enough high-end enterprises and diverse commercial businesses to boost housing prices and enhance functional diversity.

CAA03 and CAA05, located on the western bank of the Huangpu River in Huangpu District, and CAA04, directly opposite it in Pudong New District, were the areas where the landscape catalytic effect was most pronounced within the research site. Both had significant catalytic effects in the spatial aspect. In addition, CAA03 and CAA05 were more affected in the economic aspect and less so in the social aspect, while with CAA04 it was vice versa. CAA03 and CAA05 are where the Bund and the Exotic Building Clusters in the Bund of Shanghai are located. Thus, the reason these areas were weak in the social aspect was the shortage of green spaces. CAA04 is the Lujiazui International Financial Center, which was constructed long before the Huangpu River regeneration. It had the highest development density and rental price within the research site. The area is predominantly occupied by financial enterprises; thus, it has relatively low functional diversity and a smaller residential population.

The landscape catalytic effects seen in CA06 and CAA06 were generally moderate, and the impacts were relatively balanced across the three aspects. These areas are composed of high-end residential neighbourhoods well equipped with public services, infrastructures and parks, as they were developed during the boom of the Lujiazui International Financial Center. Due to the privacy of the high-end residential areas and a certain degree of exclusivity, the social effects were slightly poorer than the other two aspects in these two areas.

The 07 and 08 districts, located on both sides of the Huangpu River, were constructed in tandem with the regeneration process; thus, they appeared to be evenly developed in the three aspects and had a medium level of landscape catalytic effect. CA08 and CAA08 are where the large-scale Shanghai Expo Park is located, which led to slightly higher ratings in parks and user concentration but lower functional diversity.

To conclude, in addition to the catalyst's attributes in relation to physical form, origin, impact and first points [75], the depth and focus of landscape catalytic effects are also influenced by many contextual factors, including the original dominant functions of the catalytic area, its development conditions, spatial resources left that can be catalysed, and the timing and duration of the catalytic effect.

4.3. Comparing Huangpu River Waterfronts with Similar Catalytic Projects

Managing urban waterfront resources is fundamentally distinct from managing coastal areas in rural or remote regions, as well as sea and wetland waterfronts [76]. The revitalisation of urban waterfronts stands as one of the defining urban design and planning narratives of the late twentieth century [77] and also sites of substantial transformation, holding immense potential to attract investment and counteract local economic decline. After decades of development, waterfront regeneration is observed with both opportunities and challenges. On one hand, it has the potential of bringing the water back to cities, creating high-quality development with a vibrant mix of buildings and activities, and enhancing the identity of cities [78]. On the other hand, it may also lead to developments that feel alienated, being located at the urban periphery [77], and the erosion of cultural and historic identity as the old is replaced by the new.

Renowned waterfront regeneration projects worldwide have, to varying degrees, encountered these challenges. Some waterfront areas have successfully navigated through them, emerging as the city's only great public space—the Darling Harbour in Sydney; others, however, continue to struggle. For example, the redevelopment of London Docklands has been criticised as having failed to achieve its development objectives both physically and socially [79,80], in which planning and design intentions were subverted by concerns of power and capital [28]. Dilemmas of this kind are also observed in Cardiff and Liverpool in the UK, Baltimore and Boston in the US [27,29] and Duisburg in Germany. In the regeneration process, private enterprises held too much sway, with all parties locked in a battle for economic benefits. The government relinquished control over the quality of the public realm, failed to provide adequate infrastructure support, and overlooked the social impact. In addition, during these prolonged processes of negotiation and competition, the goals of waterfront regeneration were also constantly changing. The overall process lacked top-level guidance, resulting in conflicts between development, planning, design and management. These significantly extended the time required to complete waterfront regeneration and also increased the risks associated with the success of the regeneration. Additionally, some cases started with property-led development in the early regeneration stage. This not only led to the loss of the place identity of the waterfront area, but also the neglect of public services failed to establish a good reputation for the regenerated waterfronts.

The regeneration of the Huangpu River waterfront has clearly drawn on extensive global experience. Firstly, unlike coastal waterfronts that can easily become disconnected from the mainland, the Huangpu River enjoys a geographical advantage by running through the urban centre of Shanghai. Secondly, the regeneration is fully led by the government, with private enterprises participating only as partners. This ensures the consistency of regeneration goals and the rapid progress of the process. Thirdly, similar to Darling Harbour [31], the entire core waterfront area has forsworn economic gains, transforming into a linear public space. This transformation was essentially completed within two years, providing a solid foundation for the subsequent regeneration of surrounding areas. Fourthly, original industrial sites have been preserved and converted into museums, exhibition halls and other public cultural facilities. This not only enhances the sense of place but also creates a strong cultural and historical identity for the waterfront. Lastly, the waterfront has been segmented and planned according to the characteristics of different districts in Shanghai, highlighting unique landscape features of each section to avoid unbalanced development. Although it is still difficult to definitively determine the success of the Huangpu River waterfront regeneration at present, the landscape catalytic effects it has generated can already be effectively quantified and reflected. As the regeneration process of the Huangpu River continues to deepen, relevant research will also continue to evolve, dynamically evaluating its regeneration.

5. Conclusions

The Huangpu River waterfront is composed of a series of connected and well-designed open spaces that together constitute a giant linear landscape catalyst in the centre of Shanghai. With many contributions having been made to identify its vitality after regeneration, very limited attention has been paid to its catalytic effect on itself and its surrounding areas. Fourteen indicators were included in measuring the catalytic effects generated by the Huangpu River waterfront regeneration. These effects were also visualised for comparisons across different sections within the waterfront and across different catalytic aspects. The underlying reasons for bringing about the spatial, social and economic changes were also discussed, taking the original functions, development limitations and available resources of these areas into consideration.

Although the landscape catalytic effect of the Huangpu River waterfront regeneration has been indicated, this study has some limitations regarding data collection and analysis. First, the dataset used in this study was from 2021, the most recent year for which all data were complete and up to date. Undoubtedly, the current landscape catalytic effect is different from the situation in 2021. Moreover, the rental price data do not cover all the CAs and CAAs because of local development restrictions. Thus, the economic effect could not be comprehensively described and compared. In addition, the night light index was selected to illustrate user concentration given the data availability, but day and night scenarios may be different. The night light index can hardly indicate daytime human activities. Incorporating real walking data of people promenading along the Huangpu River during the daytime is necessary to fully depict the social effects facilitated by the regeneration in future investigations. Further explorations should also pay attention to the temporal changes of the landscape catalytic effect if data availability can meet the requirements to gain an integrated understanding of the catalytic rate in different areas, the catalytic degree across the spatial, social and economic aspects, and the duration of the effects in different areas. Another direction that requires academic attention is which attributes of a catalyst are more likely to trigger the catalytic effect and how these effects can be manipulated through the planning and design of the landscape catalyst.

The major outcome of this study is a tool for visually presenting the catalytic effects of landscape-oriented urban regeneration. This basic framework has the potential to dynamically reveal the depth and progress of a catalytic effect. Its application in this study also confirms its capacity to evaluate regeneration projects characterised by large-scale, extensive linear form, complicated surrounding contexts and long-term consecutive development. In addition, the research provides strong evidence supporting the feasibility of transforming large-scale industrial lands into urban open spaces, and also strategic implications for the decision-making process in the urban regeneration agenda that centres around waterfronts.

Supplementary Materials: The following supporting information can be downloaded at: https://www.mdpi.com/article/10.3390/land14020422/s1.

Author Contributions: Conceptualization, Y.Y. and X.X.; methodology, Y.Y. and X.X.; software, D.M. and X.X.; validation, Y.Y.; formal analysis, Y.Y. and D.M.; investigation, Y.Y.; resources, Y.Y.; data curation, Y.Y. and X.X.; writing—original draft preparation, Y.Y.; writing—review and editing, Y.Y.; visualization, D.M.; supervision, Y.Y. and X.X.; project administration, Y.Y. and X.X.; funding acquisition, Y.Y. and X.X. All authors have read and agreed to the published version of the manuscript.

Funding: This research was funded by Shanghai Key Laboratory of Urban Regeneration and Spatial Optimization Technology Open Project (20240201); Restorative Urbanism Research Center (RURC), Joint Laboratory for International Cooperation on Eco-Urban Design, Tongji University (CAUP-UD-06).

Data Availability Statement: Publicly available datasets were analyzed in this study. These data can be found here: https://zenodo.org/records/11397015 (accessed on 13 October 2024), www.openstreetmap.org (accessed on 5 November 2024), http://59.175.109.173:8888/app/login.html (accessed on 1 November 2024), https://s.weibo.com (accessed on 12 October 2024), https://www.amap.com/ (accessed on 25 October 2024), www.lianjia.com (accessed on 13 October 2024), https://earthexplorer.usgs.gov (accessed on 15 October 2024), http://geodata.nnu.edu.cn (accessed on 19 October 2024).

Conflicts of Interest: Author Dongbo Ma were employed by the Shanghai Tongji Urban Planning and Design Institute, Co., Ltd. The remaining authors declare that the research was conducted in the absence of any commercial or financial relationships that could be construed as a potential conflict of interest. Shanghai Tongji Urban Planning and Design Institute, Co., Ltd. partially financed the project. The funders had no role in the design of the study; in the collection, analyses, or interpretation of data; in the writing of the manuscript, or in the decision to publish the results.

Abbreviations

The following abbreviations are used in this manuscript:

CA	The catalyst area
CAA	The area affected by the catalyst area

References

1. Marshall, R. *Waterfronts in Post-Industrial Cities*, 1st ed.; Taylor & Francis: Abingdon, UK, 2001.
2. Cerreta, M.; Salzano, I. 'Green Urban Catalyst': An Ex Post Evaluation of Sustainability Practices. In Proceedings of the Real Corp, Sitges, Spain, 22–25 April 2009.
3. Moss, M.L. An agenda for the urban coastal zone. *Coast. Manag.* **1977**, *3*, 213–215. [CrossRef]
4. Rossi, A. *The Architecture of the City*, 1st ed.; The MIT Press: Cambridge, MA, USA, 1984.
5. Attoe, W.; Logan, D. *American Urban Architecture: Catalysts in the Design of Cities*, 1st ed.; Univ of California Press: Berkeley, CA, USA, 1989.
6. Sternberg, E. What makes buildings catalytic? How cultural facilities can be designed to spur surrounding development. *J. Archit. Plan. Res.* **2002**, *19*, 30–43.
7. Raco, M. Assessing the discourses and practices of urban regeneration in a growing region. *Geoforum* **2003**, *34*, 37–55. [CrossRef]
8. Oswalt, P.; Overmeyer, K.; Misselwitz, P. *Urban Catalyst: Strategies for Temporary Use*; Birkhäuser: Basel, Switzerland, 2011.
9. Catalyst, U. Patterns of the unplanned. In *Loose Space: Possibility and Diversity in Urban Life*, 1st ed.; Franck, K.A., Stevens, Q., Eds.; Routledge: London, UK, 2006; pp. 271–288.
10. Jacobs, J. *Death and Life of Great American Cities, Reissued ed.*; Vintage: New York, NY, USA, 1992.
11. Norberg-Schulz, C. *Genius Loci: Towards a Phenomenology of Architectures*, 1st ed.; Rizzoli: New York, NY, USA, 1979.
12. Lynch, K. *Reconsidering the Image of the City*, 1st ed.; Springer: New York, NY, USA, 1984; pp. 151–161.
13. Davis, J. Urban catalysts in theory and practice. *ARQ Archit. Res. Q.* **2009**, *13*, 295–306. [CrossRef]
14. Walkowiak, E.; Frazier, D. Brownfields redevelopment as a catalyst for creating sustainable cities. *WIT Trans. Ecol. Environ.* **2000**, *39*, 10.
15. Millspaugh, M.L. Waterfronts as catalysts for city renewal. In *Waterfronts in Post-Industrial Cities*, 1st ed.; Marshall, R., Ed.; Taylor & Francis: Abingdon, UK, 2001; pp. 82–93.

16. Allen, R.C. *The British Industrial Revolution in Global Perspective*, 1st ed.; Cambridge University Press: Cambridge, UK, 2009.
17. Wang, T. Railway station complex and urban catalyst—Example of shanghai south railway station. *Urban Plan. Form* **2006**, *4*, 76.
18. Mahmoud, W.S.; Mohammed, T.A. The Catalyst Historic Building in Urban Development. *J. Eng. Technol.* **2015**, *33 Pt A*, 1–7.
19. Healey, P.; Davoudi, S.; O'Toole, M.; Raisanaglu, S.; Usher, D. *Rebuilding the City, Property Led Urban Regeneration*, 1st ed.; Chapman & Hall: London, UK, 1992.
20. Chen, Y.; Jin, G.; Xu, Z. Study on the Catalyst Effect of Urban Complex Against the Background of Rapid Urbanisation. *Urban Plan. Int.* **2011**, *26*, 97–104.
21. Wang, Z.; Zhu, J. Quantitative Assessment of Landscape Catalytic Effect of Urban Open Space: A Case Study of Chongqing City. *Chin. Landsc. Archit.* **2024**, *40*, 77–83.
22. Turnagöl, D. Reviving the Culture and the Identity of Istanbul: A Creative Hub as a Catalyst on the waterfront. Master's Thesis, Politecnico Milano, Milano, Italy, 2019.
23. Shamsuddin, S.; Latip, N.A.; Sulaiman, A. Waterfront regeneration as a sustainable approach to city development in Malaysia. *WIT Trans. Ecol. Environ.* **2008**, *117*, 45–54.
24. Basu, D. Waterfront public spaces in port cities: A case study of Lisbon and Barcelona. Master's Thesis, Politecnico Milano, Milano, Italy, 2011.
25. Hall, P.; Tewdwr-Jones, M. *Urban and Regional Planning*, 6th ed.; Routledge: London, UK, 2019.
26. Elkadi, H.; Biscaya, S.A. Catalyst Approach for Smart Ecological Urban Corridors at Disused Waterways. *Urban Plan.* **2023**, *8*, 1–19.
27. De Jong, W.M. Revitalizing the urban core waterfront development in Baltimore, Maryland. In *Urban Regeneration in a Changing Economy: An International Prospective*, 1st ed.; Fox-Przeworski, J., Goddard, J., Jong, M., Eds.; Clarendon Press: Oxford, UK, 1991; pp. 185–198.
28. Clark, M. Fallow land in old docks: Why such a slow take-up of Britain's waterside redevelopment opportunities? *Marit. Policy Manag.* **1985**, *12*, 157–167. [CrossRef]
29. Wang, C. *Waterfront Regeneration*; Town & Country Planning Summer School, Cardiff University: Wales, UK, 2002.
30. Martí, P.; Mayor, C.G.; Melgarejo, A. Waterfront landscapes in Spanish cities: Regeneration and urban transformations. *WIT Trans. Built Environ.* **2018**, *179*, 45–56.
31. Lai, L.W.; Lorne, F.T. Sustainable urban renewal and built heritage conservation in a global real estate revolution. *Sustainability* **2019**, *11*, 850. [CrossRef]
32. Giblett, G.; Samant, S. A review of urban waterfront regeneration in global Asian port cities and the sustainability of their development. *J. Urban Regen. Renew.* **2012**, *5*, 266–279. [CrossRef]
33. Giovinazzi, O.; Moretti, M. Port cities and urban waterfront: Transformations and opportunities. *TeMA-J. Land Use Mobil. Environ.* **2009**, *3*, 57–64.
34. Knoll, M.; Lubken, U.; Schott, D. *Rivers Lost, Rivers Regained: Rethinking City-River Relations*, 1st ed.; University of Pittsburgh Press: Pittsburgh, PA, USA, 2017.
35. Gehl, J. *Life Between Buildings*, 1st ed.; Van Nostrand Reinhold: New York, NY, USA, 1987.
36. Li, M.; Shen, Z.; Hao, X. Revealing the relationship between spatio-temporal distribution of population and urban function with social media data. *Geo J.* **2016**, *81*, 919–935. [CrossRef]
37. Mehta, V. Small businesses and the vitality of main street. *J. Archit. Plan. Res.* **2011**, *28*, 271–291.
38. Unt, A.L.; Bell, S. The impact of small-scale design interventions on the behaviour patterns of the users of an urban wasteland. *Urban For. Urban Green.* **2014**, *13*, 121–135. [CrossRef]
39. Gaoalafe, P.G. Vitality in Urban Waterfront Public Spaces: The Case of Laguna, Famagusta. Master's Thesis, Eastern Mediterranean University (EMU)-Doğu Akdeniz Üniversitesi (DAÜ), Famagusta, Cyprus, 2017.
40. Yavuz, A.; Acar, C. Functional Quality of Place as Permeability Parameters at Waterfronts. *Oxid. Commun.* **2016**, *39*, 560–577.
41. Da, T.; Li, Y. Research on social connectivity of urban waterfront park based on place attachment. *J. Nanjing For. Univ.* **2023**, *47*, 227.
42. Fan, Z.; Duan, J.; Luo, M.; Zhan, H.; Liu, M.; Peng, W. How did built environment affect urban vitality in urban waterfronts? A case study in Nanjing reach of Yangtze River. *ISPRS Int. J. Geo-Inf.* **2021**, *10*, 611. [CrossRef]
43. Ding, J.; Luo, L.; Shen, X.; Xu, Y. Influence of built environment and user experience on the waterfront vitality of historical urban areas: A case study of the Qinhuai River in Nanjing, China. *Front. Archit. Res.* **2023**, *12*, 820–836. [CrossRef]
44. Hu, C.; Xu, L.; Cai, X.; Tian, D.; Zhuang, S. Research on the Spatiotemporal Pattern and Influencing Mechanism of Coastal Urban Vitality: A Case Study of Bayuquan. *Buildings* **2024**, *14*, 2173. [CrossRef]
45. Liu, S.; Lai, S. Influence Factors of Urban Public Space Vitality Based on Multi-source Data: A Case Study of Huangpu River Waterfront Area of Shanghai. *Landsc. Archit.* **2021**, *28*, 75–81.
46. Chen, L.; Ma, Y. How do ecological and recreational features of waterfront space affect its vitality? developing coupling coordination and enhancing waterfront vitality. *Int. J. Environ. Res. Public Health* **2023**, *20*, 1196. [CrossRef]

47. Liu, S.; Lai, S.; Liu, C.; Jiang, L. What influenced the vitality of the waterfront open space? A case study of Huangpu River in Shanghai, China. *Cities* **2021**, *114*, 103197. [CrossRef]
48. Niu, Y.; Mi, X.; Wang, Z. Vitality evaluation of the waterfront space in the ancient city of Suzhou. *Front. Archit. Res.* **2021**, *10*, 729–740. [CrossRef]
49. Yin, Y.; Shao, Y.; Lu, H.; Han, Y. Environmental drivers of the vital urban coastal zones: An explorative case study based on the data-driven multi-method approach. *Front. Ecol. Evol.* **2022**, *10*, 962299. [CrossRef]
50. Wang, Z.; Li, H.; He, D.; Li, X. Analyzing Urban Waterfront Spatial Openness Characteristic from Multi-Dimensional Perspective: A Case Study of Shanghai Waterfront Central District. *Plan. Constr.* **2022**, *11*, 72–78.
51. Desfor, G.; Jorgensen, J. Flexible urban governance. The case of Copenhagen's recent waterfront development. *Eur. Plan. Stud.* **2004**, *12*, 479–496. [CrossRef]
52. Long, Y.; Huang, C. Does block size matter? The impact of urban design on economic vitality for Chinese cities. *Environ. Plan. B Urban Anal. City Sci.* **2019**, *46*, 406–422. [CrossRef]
53. Sairinen, R.; Kumpilainen, S. Assessing social impacts in urban waterfront regeneration. *Environ. Impact Assess. Rev.* **2006**, *26*, 120–135. [CrossRef]
54. Othman, A.; Alhagla, K.; Hasan, A.E. The impact of attributes of waterfront accessibility on human well-being: Alexandria Governorate as a case study. *Ain Shams Eng. J.* **2021**, *12*, 1033–1047. [CrossRef]
55. Pettorelli, N. *The Normalized Difference Vegetation Index*, 1st ed.; Oxford University Press: New York, NY, USA, 2013.
56. Li, S.; Wu, C.; Lin, Y.; Li, Z.; Du, Q. Urban Morphology Promotes Urban Vibrancy from the Spatiotemporal and Synergetic Perspectives: A Case Study Using Multisource Data in Shenzhen, China. *Sustainability* **2020**, *12*, 4829. [CrossRef]
57. Im, H.N.; Choi, C.G. The hidden side of the entropy-based land-use mix index: Clarifying the relationship between pedestrian volume and land-use mix. *Urban Stud.* **2019**, *56*, 1865–1881. [CrossRef]
58. Da, T.; Xu, Y. Evaluation on connectivity of urban waterfront redevelopment under hesitant fuzzy linguistic environment. *Ocean. Coast. Manag.* **2016**, *132*, 101–110. [CrossRef]
59. Ministry of Housing and Urban-Rural Development of the People's Republic of China. Code for Transport Planning on Urban Road (GB 50200-1995). 1995.
60. Ministry of Housing and Urban-Rural Development of the People's Republic of China. Code for the Design of Public Park GB51192-2016. 2016.
61. Saaty, R.W. The analytic hierarchy process—What it is and how it is used. *Math. Model.* **1987**, *9*, 161–176. [CrossRef]
62. Qiao, L.; Zhang, Y.; Cao, W. Evaluation of urban river landscape design rationality based on AHP. *Water Sci. Eng.* **2008**, *1*, 75–81.
63. Min, J.; Pan, T. Landscape Evaluation of Forest Park Based on Analytic Hierarchy Process. *Math. Probl. Eng.* **2022**, *1*, 7496570. [CrossRef]
64. Srdjevic, Z.; Lakicevic, M.; Srdjevic, B. Approach of decision making based on the analytic hierarchy process for urban landscape management. *Environ. Manag.* **2013**, *51*, 777–785. [CrossRef] [PubMed]
65. Keyvanfar, A.; Shafaghat, A.; Mohamad, S.; Abdullahi, M.; Ahmad, H.; Mohd Derus, N.H.; Khorami, M. A Sustainable historic waterfront revitalization decision support tool for attracting tourists. *Sustainability* **2018**, *10*, 215. [CrossRef]
66. Huang, Q.; Huang, J.; Zhan, Y.; Cui, W.; Yuan, Y. Using landscape indicators and Analytic Hierarchy Process (AHP) to determine the optimum spatial scale of urban land use patterns in Wuhan, China. *Earth Sci. Inform.* **2018**, *11*, 567–578. [CrossRef]
67. Li, G.; Li, Y. Optimization spatial pattern method for vegetation landscape in bay based on AHP. *Microprocess. Microsyst.* **2021**, *83*, 104041. [CrossRef]
68. Balsas, C.J. Exciting walk-only precincts in Asia, Europe and North-America. *Cities* **2021**, *112*, 103129. [CrossRef]
69. Beyard, M.D.; Braun, R.E.; McLaughlin, H.; Phillips, P.L.; Rubin, M.S.; Berens, G.L. *Developing Urban Entertainment Centers*, 1st ed.; Urban Land Inst: Washington, DC, USA, 1998.
70. Wu, Z. World EXPOs: As a Catalyst for Urban Sustainability. *Time Archit.* **2008**, *4*, 11–17.
71. Grodach, C. Museums as Urban Catalysts: The Role of Urban Design in Flagship Cultural Development. *J. Urban Des.* **2008**, *13*, 195–212. [CrossRef]
72. Camhis, M.; Fox, S. The European Community as a catalyst for European urban networks. *Athens Cent. Ekistics* **1992**, *29*, 4–6.
73. Benneworth, P.; Dauncey, H. International urban festivals as a catalyst for governance capacity building. *Environ. Plan. C Gov. Policy* **2010**, *28*, 1083–1100. [CrossRef]
74. Kristo, S.; Dhiamandi, J. Urban Catalyst as the tool for public space transformation. In Proceedings of the 1st International Scientific Conference on Professional Sciences, Durres, Albania, 25–26 November 2016.
75. Balvočienė, V.; Zaleckis, K. Evaluation of urban catalysts through history. *Archit. Et Artibus* **2020**, *12*, 5–13.
76. Chang, S.U. The Coast, Social Actions, and Architectural Production: A Review of The Social Imperative: Architecture and the City in China. *Landsc. Archit. Front.* **2023**, *11*, 77–81.
77. Dovey, K. *Fluid City; Transforming Melbourne's Urban Waterfront*, 1st ed.; Australia UNSW Press: Sydney, Australia, 2004.

78. Shaw, B. History at the water's edge. In *Waterfronts in Post-Industrial Cities*, 1st ed.; Marshall, R., Ed.; Taylor & Francis: Abingdon, UK, 2001.
79. Brownill, S. *Developing London Docklands: Another Planning Disaster?* 1st ed.; Paul Chapman: London, UK, 1990.
80. Malone, P. *City, Capital and Water*, 1st ed.; Routledge: London, UK; New York, NY, USA, 1996.

MDPI AG
Grosspeteranlage 5
4052 Basel
Switzerland
Tel.: +41 61 683 77 34

Land Editorial Office
E-mail: land@mdpi.com
www.mdpi.com/journal/land

www.ingramcontent.com/pod-product-compliance
Lightning Source LLC
Chambersburg PA
CBHW061603210926
51850CB00009B/2195
9783725868209